T0281488

Strategic Research on Construction and Promotion of China's Intelligent Cities

Editor-in-chief

Yunhe Pan, Chinese Academy of Engineering, Beijing, China

This book series is the first in China on "Intelligent City" research, with systematic and thorough contributions from more than 200 Chinese experts including 47 academicians of the Chinese Academy of Engineering (CAE) in related fields. The book series is co-published with Zhejiang University Press, Hangzhou, China and consists of 13 volumes as planned, including one general report and 12 sector reports. In 2010, CAE conducted a research on the development of "smart cities" and concluded that urban development in China has reached a crucial turning point. Therefore, CAE kicked off the key consultancy research project on "Strategic Research on Construction and Promotion of China's Intelligent Cities", on which this book series is based. Firsthand and research results, surveys and analysis are provided on almost every aspect of urban development and smart cities in this series. Representing the highest level of research in this field in China, the book series will offer an authoritative reference resource for international readers, helping them to understand intelligent city construction in China, a movement expected to be highly influential around the globe.

More information about this series at http://www.springer.com/series/15953

Yunhe Pan

Strategic Research on Construction and Promotion of China's Intelligent Cities

General Report

ZHEJIANG UNIVERSITY PRESS
浙江大学出版社

Yunhe Pan
College of Computer Science and Technology
Zhejiang University
Hangzhou, Zhejiang
China

ISSN 2522-8943   ISSN 2522-8951   (electronic)
Strategic Research on Construction and Promotion of China's Intelligent Cities
ISBN 978-981-13-3878-6   ISBN 978-981-10-6310-7   (eBook)
https://doi.org/10.1007/978-981-10-6310-7

Jointly published with Zhejiang University Press, Hangzhou, China

The print edition is not for sale in China Mainland. Customers from China Mainland please order the print book from: Zhejiang University Press, Hangzhou, China

Printed on acid-free paper

This Springer imprint is published by Springer Nature
The registered company is Springer Nature Singapore Pte Ltd.
The registered company address is: 152 Beach Road, #21-01/04 Gateway East, Singapore 189721, Singapore

Foreword

In 2008, IBM proposed the concept of "Smarter Planet," in which "Smart City" was one of its components, mainly referred to 3I, namely instrumented, interconnected, intelligent, and the goal was to implement the company's "solutions," such as smart transportation, medical, government services, monitoring, grid, water, and other items.

In early 2009, US President Barack Obama publicly acknowledged IBM's "Smarter Planet" concept. In December 2012, the Global Trends 2030 (Atlantic Council 2013), published by the National Intelligence Council, noted that the four most influential technologies for global economic development were information technology, automation and manufacturing technology, resource technology, and health technology, in which "Smart City" was one of the information technology contents. Envisioning 2030: US Strategy for the Coming Technology Revolution report pointed out that the world was on the cusp of the next major technological change, the "third industrial revolution" represented by manufacturing technology, new energy and Smart City would have an important influence on shaping future political, economic and social development trends.

In May 2011, after the implementation of the "i2010" Strategy, the EU Net! Works forum introduced Smart Cities Applications and Requirements White Paper, emphasizing low carbon, environmental protection, and green development. After that, the EU said it would set "Smart City" as the key development content of the Eighth Framework Programme (FP8).

In August 2009, in the Smarter Planet to win in China plan, IBM tailored for the Chinese six intelligent solutions: "Intelligent Power," "Intelligent Medical," "Intelligent City," "Intelligent Traffic," "Intelligent Supply Chain," and "Intelligent Banking." In 2009, "Smart City" spreads in succession at all levels in China, as till September 2013, there were a total of 311 cities under construction or planning to build Smart City in China.

In 2010, Chinese Academy of Engineering carried out research on the "smart city" construction. It considered that the current urban development in China had reached a key transition period, but since national conditions were different, there were still some problems in "Smart City" construction.

To this end, in February 2012, the Chinese Academy of Engineering launched a major consulting research project "China Intelligent City Construction and Promotion Strategy Research." Since the project started, many city leaders and scholars have shown a keen interest, and expected to devote to the research and practice of intelligent city construction. With the strong support of people from all walks of life and the efforts of academicians and experts of the "China Intelligent City Construction and Promotion Strategy Research" project group of the Chinese Academy of Engineering, we have combined three research efforts: experts of the relevant national ministries (such as the National Development and Reform Commission, the Ministry of Industry and Information Technology of the People's Republic of China, the Ministry of Housing and Urban-Rural Development of the People's Republic of China), experts of typical cities (such as Beijing, Wuhan, Xi'an, Shanghai, Ningbo), 47 academicians and more than 180 experts of the Department of Information and Electronic Engineering of the Chinese Academy of Engineering, the Department of Energy and Mining Engineering, the Department of Environment and Textile Engineering, the Department of Engineering Management and the Department of Civil Engineering, Water Conservancy and Construction Engineering, and other departments. The research project was divided into 13 task groups, involving urban infrastructure, information, industry, management, and so on. In addition, the project also has a comprehensive group, whose main task is to comprehensively formulate the integrated volume of the book series *Strategic Research on Construction and Promotion of China's Intelligent Cities* on the basis of the results of 13 task groups.

After more than two years, the research team has formed a number of research results and research comprehensive report by carrying out inspection and investigation in the site, carrying out the forums and exchanges with experts and scholars at home and abroad, having informal discussion with the national authorities and local authorities responsible comrades, and team research and analysis itself and so on. In the study, we put forward that it will be more suitable for China's national conditions to carry out intelligent city (iCity) construction and promotion in China. The intelligent city construction will become an accelerant of deepening the system reform and development, and become a strong starting point for the economic and social development and the realization of "China Dream" in China.

Christopher Kojm

Reference

Atlantic Council, 2013. Envisioning 2030: US Strategy for the Coming Technology Revolution. Washington, D.C.: Brent Scowcroft Center on International Security.

Preface

The book series *Strategic Research on Construction and Promotion of China's Intelligent Cities* is compiled and published by 47 academicians and over 180 experts on the basis of the research achievements obtained after two years of in-depth investigation, research and analysis, and the study on China's Intelligent City Construction and Promotion Strategy, which is one of the major consulting and research projects conducted by the Chinese Academy of Engineering after revision in accordance with publishing requirements. The book series, consisting of 1 comprehensive volume and 13 sub-volumes, have been published in succession by Zhejiang University Press. The comprehensive volume mainly discusses how to conduct the intelligent city construction and promotion with Chinese characteristics in the intelligent urbanization development of our future cities; the sub-volumes focus on the construction and promotion of intelligent city in terms of urban economy, science, culture, management and education, spatial organization pattern of cities, intelligent transportation and logistics, intelligent grid and energy network, intelligent manufacturing and design, knowledge center and information processing, intelligent information network, intelligent building and smart home, intelligent medical and health care, urban security, urban environment, intelligent business and finance, intelligent city's time and space information infrastructure, intelligent city's evaluation indicator system, etc.

As a consultant of "Strategic Research on Construction and Promotion of China's Intelligent Cities" project group, I have participated in several research meetings of the project group and put forward some "humble opinions." Overall, I think, under the leadership of the project group leader—academician Pan Yunhe—"Strategic Research on Construction and Promotion of China's Intelligent Cities" has made significant progress, which is mainly shown in the following aspects.

Since the 1990s, the world entered into the era of information technology, and the city had gradually developed from the traditional binary space to the ternary space. The mentioned first metaspace refers to a physical space (P), which consists of physical environment of the city and the urban physical; the second metaspace refers to a human social space (H), that is, the human decision-making and social interaction space; the third metaspace refers to a cyberspace (C), that is, the "network information" space composed of computer and Internet. The city intelligence is the development trend of cities throughout the world. Only the development stage of cities in each country is different, and the content is different. At present, the "Smart City" construction proposed at domestic and abroad is mainly focused on the building of the third metaspace, and the city intelligence of our country should be "ternary space" to coordinate with each other so that planning and industry, life and social and social public service could be mutual cooperation and promotion; it should be beyond the existing e-government, digital city, network city, and Smart City construction concept.

The new technological revolution will promote the arrival of urban intelligence era. Nowadays, about the new technological revolution, authors holding varying viewpoints: there it is described as "the second economy", "the third industrial revolution", "industry 4.0" "the fifth industrial revolution" and other concepts. And when it comes to the city, the new technological revolution is characterized by integrating a new generation of sensor technology, Internet technology, big data technology, and engineering technical knowledge into the city's systems to form the upgrading of urban construction, urban economy, urban management, and public service, so as to embrace a new era of urban intelligence development. If China's urbanization and the new technological revolution are organically linked together, it can not only promote the benign and healthy development of China's urban intelligence process, but also promote the birth of more new technologies. China shall undoubtedly actively participate in this process and make a greater contribution to the development of the world's economy, science, and technology.

It has been repeatedly deliberated and considered by the project group to use "Intelligent City" (iCity) to replace "Smart City." The reasons are as follows: First of all, the western developed countries have completed the urbanization, industrialization, and agricultural modernization, the main tasks of the Smart City they refer to are limited to the intelligence of government management and service, and the administrative functions of their city managers are much narrower than those of city mayors in China; secondly, currently China is in the simultaneous development stage of industrialization, informatization, urbanization, and agricultural modernization, the confusions and problems it encountering have uniqueness in quality and quantity, so China's urban intelligence development path must be different from the Europe and the USA, and it will be difficult to solve many development problems which Chinese cities confront by only interpreting the Smart City from the perspective of developed countries and moving this concept to China. Thus, the project group decided to use the term "Intelligent City" (iCity) and expected this term would be more in line with China's national conditions.

The construction and promotion of intelligent cities have far-reaching significance to China's current economic and social development, the construction and promotion of intelligent cities are just located in the cross-point of "Four Modernizations," and its significance is mainly embodied in the following aspects. Firstly, it can be used as the basic platform for the simultaneous development of "Four Modernizations" and become a focal point of China's economic and social development to avoid the "middle income trap," so as to create a new urbanization development path with Chinese characteristics. Secondly, by putting the intelligent city as an important basis (point), it can promote the development of "One Belt, One Road" (line) and new area (plane) and constitute a reasonable development layout of "point, line, and plane." Thirdly, it is conducive to promote the structure upgrading and transformation of manufacturing and its service industry, to achieve the transformation from the urban industry to the intensive pattern so as to slow down the material growth, accelerate the value growth, and improve the added value; it is conducive to the usage and integration of a variety of e-commerce, big data, cloud computing, and Internet of Things technologies to achieve the "broadband, pan, mobile, integration, security, green" development of information and network technology, promote the improvement of urban industry efficiency, form the new factors of production and new formats, and create new conditions for entrepreneurship and employment. Fourthly, it developed from the simple and linear decision-making based on limited information to the networked and optimized decision-making based on urban comprehensive system information, so as to help the government improve the level of urban management services and promote the in-depth urban administrative system reform and development. Fifthly, it can use new technologies to optimize and improve the planning of urban construction, roads, transportation, energy, resources, environment, etc., to increase the utilization efficiency of elements; and further protect, inherit, develop, and sublimate the urban history, landforms, local culture, etc.; and achieve the change of the public health management from the concept to reality and so on. Sixthly, it can find and cultivate a number of urban planners, management experts, high-level scientists, data science and security experts, engineering and technical experts, etc., to adapt to the new technological revolution trends; learn from past experience and lessons, and pay attention to the renovation during intelligent city operation and maintenance, and it can focus on cultivating a large base number of operation and maintenance engineers and technical staff of urban functions to both understand the theory and practice, achieving the gradual transformation from relying on the demographic dividend to relying on knowledge and talent dividends so as to support health, sustainable development of China's urban intelligence.

To sum up, the book series *Strategic Research on Construction and Promotion of China's Intelligent Cities* has rich content and clear views, and the proposed development goals, ways, strategies, and recommendations are reasonable and operational. I think this series of books is the literature of urban management innovation and development research with high reference value, and it has important theoretical significance and practical value to the development of new urbanization in China. I believe that readers of all sectors will get a lot of new inspiration and harvest after reading them, and the book will inspire the enthusiasm of everyone to participate in the construction of intelligent city, so as to put forward more thinking and unique insights.

China is a developing country with long history and much agricultural population, and is committed to the economic society to be developed in good and fast and commercial manner and the construction of new urbanization. I am convinced that the publication of the book series *Strategic Research on Construction and Promotion of China's Intelligent Cities* will play an active and positive role in promoting this. Let us strive and work together for the realization of the great "Chinese dream"!

Hangzhou, China   Yunhe Pan
January 2015

Contents

Contributors

Project Consultant

Xu Kuangdi, Vice Chairman, The Tenth, CPPCC National Committee, Honorary Chairman of the Presidium and Academician, Chinese Academy of Engineering
Zhou Ji, President and Academician, Chinese Academy of Engineering

Project Group Leader

Pan Yunhe, Academician of the Chinese Academy of Engineering

Vice Project Group Leader

Mao Guanglie, Vice Governor, People's Government of Zhejiang Province
Liu Yunjie, Academician, China Unicom

Project Members

Xu Qingrui, Academician, Zhejiang University
Wang Yingluo, Academician, Xi'an Jiaotong University
Wang Zhongtuo, Academician, Dalian University of Technology
Zou Deci, Academician, China Academy of Urban Planning and Design
Shi Zhongheng, Academician, Beijing Urban Engineering Design and Research Institute
Yu Yixin, Academician, Tianjin University
Cen Kefa, Academician, Zhejiang University
Ni Weidou, Academician, Tsinghua University
Wu Cheng, Academician, Tsinghua University

Sun Youxian, Academician, Zhejiang University

Xu Zhilei, Academician, China Academy of Engineering Physics

Wang Tianran, Academician, Shenyang Institute of Automation, Chinese Academy of Sciences

Zhong Zhihua, Academician, Chinese Academy of Engineering

Li Bohu, Academician, the Second Research Institute of China Aerospace Science & Industry Corporation

Li Guojie, Academician, Institute of Computing Technology, Chinese Academy of Sciences

Li Youping, Academician, Beijing Institute of Applied Physics and Computational Mathematics

Li Deyi, Academician, The 61st Institute of the General Staff Headquarters of Chinese People's Liberation Army

Liu Yongcai, Academician, The Third Research Institute of China Aerospace Science and Industry Corporation

Zhu Gaofeng, Academician, Ministry of Industry and Information Technology

Chen Junliang, Academician, Beijing University of Posts and Telecommunications

Wu Jiangxing, Academician, The PLA Information Engineering University

Zheng Nanning, Academician, Xi'an Jiaotong University

Jiang Yi, Academician, Tsinghua University

Sun Yu, Academician, The 54th Institute of the Ministry of Industry and Information Technology

Li Lanjuan, Academician, State Key Laboratory for Diagnosis and Treatment of Infectious Diseases, Zhejiang University

Yang Shengli, Academician, Institute for Systems Biology, Shanghai Jiao Tong University

Wang Weiqi, Academician, Fudan University

Wu Manqing, Academician, China Electronics Technology Group Corporation

Zhong Shan, Academician, The Second Research Institute of China Aerospace Science and Industry Corporation

Meng Wei, Academician, Chinese Research Academy of Environmental Sciences

Ren Zhenhai, Academician, Center for Climate Impact Research of State Environmental Protection Administration

Jin Jianming, Academician, Fudan University

Liu Hongliang, Academician, Chinese Research Academy of Environmental Sciences

Hao Jiming, Academician, Tsinghua University

Hou Li'an, Academician, The Second Artillery Engineering Design Institute

Qu Jiuhui, Academician, Research Center for Eco-Environmental Sciences, Chinese Academy of Sciences

Duan Ning, Academician, Chinese Research Academy of Environmental Sciences

Zhang Yaoxue, Academician, Central South University

Ning Jinsheng, Academician, Wuhan University
Wang Jiayao, Academician, The PLA Information Engineering University
Zhang Zuxun, Academician, Wuhan University
Li Jiancheng, Academician, School of Geodesy and Geomatics, Wuhan University
Shen Jian, Vice President, Design and Research, Zhejiang Provincial Institute of
    Communications Planning
Yu Hongyi, Deputy Secretary, Ningbo Municipal Committee
Li Renhan, Inspector (Chair), The 3rd Institute of the Chinese Academy of
    Engineering
Wu Zhiqiang, Vice President, Professor, Tongji University

Project Author Group

Li Renhan, Inspector (Chair), The 3rd Institute of the Chinese Academy of
    Engineering
Wu Zhiqiang, Vice President, Professor, Tongji University
Chen Jin, Professor, Tsinghua University
Ceng Yuan, Professor, Tianjin University
Yu Zitao, Professor, Zhejiang University
Yao Haipeng, Lecturer, Beijing University of Posts and Telecommunications
Wang Fulin, Associate Professor, Tsinghua University
He Qianfeng, Intermediate Title, Zhejiang Digital Medicine and Health Technology
    Institute
Gu Xinjian, Professor, Zhejiang University
Deng Chao, Professor, Central South University
Chen Bo, Deputy Director, Ningbo Academy of Smart City Development
Liu Chaohui, Researcher, Intelligent Urbanization Co-creation Center, Tongji
    University
Liu Zhi, Assistant to the President, Public Safety and Technology Institute
Hong Xuehai, Researcher, Chinese Academy of Sciences, Institute of Computing
    Technology
Huang Tao, Associate Professor, Beijing University of Posts and
    Telecommunications
Pan Liwei, Deputy Director, Associate Researcher, Center for Public Safety System
    Integration Engineering, The 38th Institute of the China Electronics Technology
    Group Corporation
Yan Li, Professor, Wuhan University
Gao Xingbao, Researcher, Chinese Research Academy of Environmental Sciences
Liu Xiaolong, Engineer, Consulting Service Center, Chinese Academy of
    Engineering
Tian Yun, Postdoctorate, Chinese Academy of Engineering

Members of Project Office

Li Renhan, Inspector (Chair), The 3rd Institute of the Chinese Academy of Engineering

An Yaohui, Deputy Director General, The 3rd Institute of the Chinese Academy of Engineering

Gao Zhanjun, Deputy Inspector, The 3rd Institute of the Chinese Academy of Engineering

Fan Guimei, Deputy Director, The 3rd Institute of the Chinese Academy of Engineering

Teng Yinglei, Associate Professor, Beijing University of Posts and Telecommunications

Chen Bingyu, Principal Staff Member, The 3rd Institute of the Chinese Academy of Engineering

Yang Yi, Teaching Assistant, The 3rd Institute of the Chinese Academy of Engineering

Hu Nan, Principal Staff Member, General Office, Chinese Academy of Engineering

Chapter 1
A Global Review on Smart City Development

1 A Review on Smart City Development Abroad

Before the proposal of Smart City, all countries and regions around the world were carrying out the construction of Digital City and Wireless City. In the late years, they gradually converted to the building of Smart City, which actually reflects a changing from digitalization to intelligence.

1.1 USA

1. Status Quo

After the implementation of the National Information Infrastructure (NII) and the Global Information Infrastructure (GII) program, American President Barack Obama publicly acknowledged the concept of Smarter Planet put forward by IBM in early 2009. In December 2012, the Global Trends 2030, published by the National Intelligence Council, noted that the four most influential technologies for global economic development are information technology, automation and manufacturing technology, resource technology, and health technology, in which "smart city" is one of the contents of information technology.

It is worthy of noting that after the release of Vision 2030: *America's Strategies in Post-western World*, this series report of about Strategic Prospect Initiatives at the year's end of 2012, Brent Scowcroft International Security Research Center under the American Atlantic Ocean Council launched the report, *Vision 2030: the Strategy of America's Response to the Next Technology Revolution* co-compiled by Mathew Burrows, Robert Manning and so on (Atlantic council 2013). It pointed out that the world is on the cusp of coming the next round of major technological reform and this 'third industrial revolution' represented by manufacturing technology, new energy and smart city would will exert major clout in reshaping the

© Springer Nature Singapore Pte Ltd. and Zhejiang University Press, Hangzhou 2018
Y. Pan, *Strategic Research on Construction and Promotion of China's Intelligent Cities*,
Strategic Research on Construction and Promotion of China's Intelligent Cities,
https://doi.org/10.1007/978-981-10-6310-7_1

future of political, economic and social development trends It recommended that the American government top up research and development expenditure to maintain the leading edge in the science and technology sector.

In recent years, the US government is deploying financial funds to boost the infrastructure construction in some of the key smart cities; through a variety of fiscal and financial policies to guide enterprises, universities and colleges, as well as various research institutes to act as the main body to make innovation in businesses and industrial model; I through the governmental outsourcing, procurement from companies, and attracting companies to get involved in the construction and operation to build smart cities with Internet, Internet of Things, broad band and other network mix as the base. The construction of Information infrastructure, intelligent grid, intelligent transportation, smart medical and other construction are the key points of current smart construction in the US.

In terms of information infrastructure construction, Obama has incorporated broadband network construction into the national policy level in January, 2009. In March, 2010, FCC officially announced the next ten years development plan of the US high-speed broadband network, which will increase the broadband speed by 25 times, which make the average speed of Internet transmission for 100 million households improving from 4 to 100 Mbps.

In terms of smart grid construction, Boulder city in Colorado launched the intelligent grid city project in August 2008, becoming the first city ever carried out intelligent grid construction. In February 2009, the United States released the "Economic Recovery Plan", intended to invest 11 billion US dollars to build a new generation of intelligent grid which a variety of control equipment would be installed. In June the same year, the US Department of Commerce and the Department of Energy jointly issued the industry standards for the first batch of intelligent grid, marking the US intelligent grid project was officially launched.

Intelligent transportation construction mainly involves traffic monitoring, intelligent control of traffic signals, non-stop charge, vehicle infrastructure integration and autonomous vehicles, intelligent traffic management system as well as other fields, which basically realize the functional support for traffic security, avoidance of traffic accidents, provision of accident rescue and speedy restore of traffic order in accident scenes.

In terms of smart medical construction, in 2002, America advocated escalation of information technology building for medication to make sure that bulk of Americans own electronic record on health in the upcoming decade in 2004, it initiated 'all-people e-archive project on health' and gradually set up National Health Information Network (NHIN). It also built regional health information network in different regions. In 2009, Obama explicitly proposed electronization on health archive of all American citizens and announced that it would input 20 billion dollars for development of information technology system of health archive. Use of internet-sharing e-archive to replace archives by paper helps boost medication efficiency, bring down errors of repetitive diagnosis and medication, reduce medication cots and attain genuine sharing of health information among medication outfits and mutual consistency among different systems.

2. **Development Trend**

Accurate and reliable city remote network. With internet of things, internet, broadband network and cloud computation interweaving, accurate, visible, reliable and intelligent city operation management network can cover all city factors and validly back up secured and reliable operation of cities.

Virtual, individualized and handy life style for residents. The new generation of intelligent information infrastructure would serve as a portal for all people to get access to the internet, other people or things anytime and anywhere. Remote medication, remote education, digital entertainment and other internet public services help optimize people's study, work and living environment.

Booming digital economy and intelligence industry. Creation and use of knowledge, intensive new and high-tech industry of knowledge and technology and modern service industry form most pivotal pillar sectors in smart cities.

Efficient and transparent governmental management. Open data, high efficiency, transparency and timely and seamless public services will become major characteristics of American government. The governmental organization will be further flattened and the business chain will attain integrated management, which would largely boost administration efficiency, bring down administration costs and further spur public participation to the process of democratic governance.

1.2 EU

1. **Status Quo**

In order to boost advancement of smart city in Europe, EU is putting into force initiatives to spur development of smart city by three stages, i.e., 'i2010' strategies (European Commission 2005), Europe 2020 strategies and 'smart city and community, European innovative partner initiatives'. Under the overall planning, it would progressively advance and assist membership countries with development of smart cities. The EU Commission has put information and communication technology as strategic development keys for 2010 in Europe and enacted *Roadmap on Research of Internet of Things Strategies*. In 2007–2013, EU input around 2 billion euros to research and development of information and communication technologies. Net Works Forum of EU launched Intelligent Cities Applications and Requirements white paper in May, 2011 that stressed low carbon, environmental and green development. After that, EU alleged that it wanted to deem 'intelligent city' as major development content for the 8th Scientific Framework Designing (FP8).

Smart city construction in Europe adopted the form of cooperation between government and companies. The government would carry out unified planning and organization while the company gets actively involved in co-launching smart city construction. Its organization models mainly cover multiple forms such as governmental investment management, participation of research outfits and non-profit outfits, public-private joint venture building and management, investment and development in telecommunication companies and so forth that reflected diversified development under the overall framework as well as characteristics combining natural talent with human activities.

2. **Development Trend**

Smart city construction in Europe closely combines city information system with economic development, city management and public services, optimizes city management decisions and technology innovation, extends industrial space and roundly boost city life quality. Extensive public participation and top-to-bottom information response mechanism can spur high fusion of city construction with the society, makes economic and social development intelligent and sustainable. It mainly contains the following five points: it will forge ubiquitous network, intensify smart city network infrastructure building, stress green information and communication technology, attain transformation of low-carbon economy, put internet of things as a pivotal round for smart city construction in EU, encourage public participation, stress social infrastructure construction, forge a hospitable social surrounding and boost degree of network of public life and public service level in cities.

1.3 Japan

1. **Status Quo**

Japan's effort to implement 'u-Japan'[1] and 'i-Japan'[2] Initiatives are aimed at incorporating digital information technology to all corners of life. For the moment, it concentrates the goal on electronic governmental governance, medication health and information services, education and talent cultivation in the hope that construction and transform of city intelligence can reshape the whole economy and society, spur new vigor and realize active and independent innovation.

---

[1]"u"stands for the English word "ubiquitous". This Initiative was designed to hasten the birth of the new revolution of information technology and to realize a accessible society where convenience is ubiquitous.

[2]"i-Japan" means the common application of digital technologies in the society of Japan and the realization of innovative society by the year 2015.

2. Development Trend

In Copenhagen Manifesto of COP 15 under *United States Framework Convention on Climate Change* in September, 2009, the Japanese government promised that it would realize the intermediate goal of bringing down reduction by 25% in 2010. In June, 2010, the Japanese Cabinet passed adjustment of three strategies for future industrial structure namely *Prospects of Industrial Structure, 2020, New Growth Strategies* and *Overall Planning on Resources*. The three strategies all took renewable energy, resource sector, environmental protection industry and intelligent power grid as major orientations. In city construction in Japan, intelligent application construction will be further intensified pertinent to different sectors as clothing, dining, travelling and accommodation.

1.4 South Korea

1. Status Quo

The South Korean government initiative smart city construction with Seoul as representative in 2006. The planning is called 'u-City'. It attempted to realize that consumers can handily develop remote education, medication, transaction of taxes and intelligent monitoring on energy consumption in home buildings via integration of public communication platforms and ubiquitous network connection. It emphasizes cultivation and development of strategic emerging industry with internet of things, biological chips, NT and other emerging u technology as the backup as new impetus for economic sustainable development. In terms of internet of things. It passed *Basic Planning on Infrastructure Building of Internet of Things* and determined that the market of internet of things was taken as a new growing momentum. In terms of cloud computation, the government launched *Comprehensive Planning on 'Cloud Computation Activation'* and planned to input 614.6 billion won prior to 2014. It intended to grow into the most powerful country in global cloud computation.

2. Development Trend

Its main contents cover in nearly a decade: work on rules, further propel standardized development of u industry, work on standards and appraisal system, further boost u-city planning and get ready for comprehensive development planning of u-City in the second stage (2014–). Breaking of information barriers and transforming of 'old city town' with u-City technologies also form quagmires to be redressed by u-City Initiatives. U-City is diversifying the connotation of u-City and proposing that u-City is amounting to urban planning + intelligent city.

Japanese and South Korean Mode: shared model for smart city construction in Japan and South Korea is u-Japan and u-Korea. It aims at bringing out 'ubiquitous' network society. The two countries show similarities in main body, strategy, route and measures of smart city construction. In terms of building main bodies, the government is responsible for coordination and the company takes the charge; in terms of building strategies, it adopts overall planning and promotion by stages; in terms of building routes, it escalates steadily and makes orderly connection; and in terms of building measures, it adopts technological problem-addressing and industrial advancement.

Measured from smart city construction abroad, America tops all countries in attaching importance with Europe following it. America and European countries put long-term city information (digital city) building as the basis, combines with latest results of information technology and consults to such means as promotion of development concepts, guidance of long-term strategic planning, construction of development model, innovation and application of information technology, supporting of laws and policy building and so forth to spur smart city building. In actuality, they also propel advancement of city intelligence.

2   Development Status of Intelligent City Construction in China

2.1   Huge Achievements on Urbanization and Information Construction in China

For over three decades, city development in our country has obtained astounding achievements that roots good foundation for city development in the future. The following achievements are obtained.

1. **Leap-frog development of urbanization**

Since 2000, demographic urbanization rate in our country rose by around 1.36% on the yearly basis and showed an annual average growth of around 21 million city residents. In 2013, urbanization ratio in our country reached 53.73% and per capita GDP reached 6807 dollars. Following the condition of urbanization level at 70% in advanced countries, China will witness around 0.1 billion people relocated to cities in the upcoming decade. For the moment, China has 288 municipal cities including 14 cities with population tallying over 4 million. Its city water use popularity ratio reaches 97% and popularity of urban gas reached 92.4%. 11.8 vehicles are owned per 10,000 people, per capita road reaches 13.8 m^2, per capital garden green area is

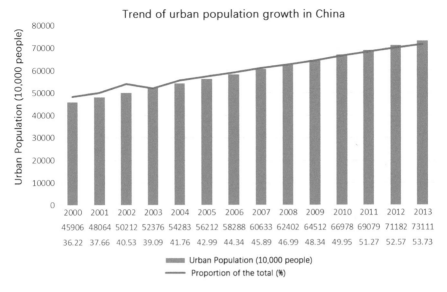

Fig. 1 Growth of city population in China (*source* China Statistical Bureau)

11.8 m² and 2.9 public toilets are owned per 10,000 people. So it has obtained great achievements.

But after going through years-long speedy development, we have fully acknowledged gigantic pressure imposed by demographics (Fig. 1), environment (Fig. 2), resources (Table 1), traffic (Fig. 3) and all aspects of the society. The unsustainable property of the current development mode has become consensus by the government, all walks of the society and the people.

2. Leap-frog advancement of informatization

The development level of information has become one of the major indicators deciding development level of productivity in China and weighing a country's comprehensives state power and international competitiveness. In recent years, informationized development speed has been gathering space. In 2011, informationized development index[3] reached 0.732 (see Fig. 4). The annual average growth rate in 2000–2011 reached 3.64%. In recent years, informationization has been showing great progress that has been making great contributions to both domestic and globalized informationization development.

First of all, competitiveness of internet communication industry has been continuously enhanced. In 2013, Huawei and Zhongxing surpassed 30% of shares in

---

[3]Informatization Development Index (IDI) measures and reflects the general level of the informatization development in a state or a region, from the aspects of information infrastructure construction, information application standards and constraint environment, and the information consumption level of the residents, etc.

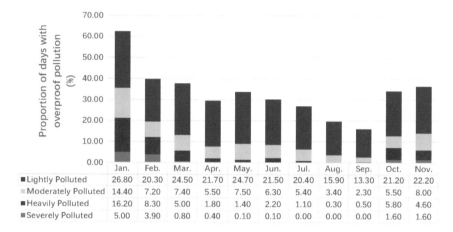

	Jan.	Feb.	Mar.	Apr.	May.	Jun.	Jul.	Aug.	Sep.	Oct.	Nov.
Lightly Polluted	26.80	20.30	24.50	21.70	24.70	21.50	20.40	15.90	13.30	21.20	22.20
Moderately Polluted	14.40	7.20	7.40	5.50	7.50	6.30	5.40	3.40	2.30	5.50	8.00
Heavily Polluted	16.20	8.30	5.00	1.80	1.40	2.20	1.10	0.30	0.50	5.80	4.60
Severely Polluted	5.00	3.90	0.80	0.40	0.10	0.10	0.00	0.00	0.00	1.60	1.60

Fig. 2 Chart of 74 cities (Beijing, Tianjin, Hebei, Yangtze River Delta Area, Pearl River Delta Area, municipality directly under the central government, provincial capitals and municipalities with independent planning status) flunking air-quality test (*data source* China Environment Monitoring Committee)

global market of communication devices. Huawei's core router is expected to break the long-term monopoly by Cisco, and the number of domestic independent patent accounted for 40% in TD-LTE patent. With the robust of domestic smart phones, shipments of domestic branded smart phones in 2013 raised to 0.324 billion, with a year on year (YoY) growth of 71.6%, and that accounted for 76.6% of smart phone's total shipments. In contrast to the long-drawn history that domestic brands took to be prevailed in the functional phone market, domestic brands of smart phones greatly shortened the stage of competition and took only two years to establish the local advantages.

Second, infrastructures such as optical broadband and mobile internet have become hotspots for informationization building. While the world as a whole intensifies information communication infrastructure building, our country is vigorously launching a string of policy deployment on broadband strategies to spur access to the new stage of basic informationizaiton building. By November, 2013, the number of users with high-speed broadband at 4 Mbps and above has reached 77.4%, 3G network had covered all rural and urban areas, trial on TD-LTE expansion of scale was successfully carried out and 4G commercialization was roundly launched.

Third, the new generation of information technology has become a major engineer for synchronous development of 'four modernization'. In recent years, domestic scale on internet of things remained a rapid growth at 30%. Remote control system has become a standard configuration in such sectors as steel metallurgy, petrol and petrochemical, chemical equipment manufacturing, logistics and other sectors. Through integration of industrial resources, 'industrial cloud' timely upgrades industrial software and deployment of information sources in the cloud

Table 1 Statistics on resource-exhausted cities announced by the country

Province (district, city)	The first batch of 12 (2008)	The second batch of 32 (2009)	The third batch of 25 (2011)	For the forest region of Great Khingan and Lesser Khingan mountains, refer to nine cities enjoying policy privilege
Hebei		Lower Garden district Yingshouyingzi Mining district	Jingxing mining area	
Shanxi		Xiaoyi city	Huozhou city	
Inner Mongolia		Arxan city	Wuhai city Shiguai district	Yakeshi city Ergun city Genhe city Oroqen Banner Zhalantun city
Liaoning	Fuxin city Panjin city	Fushun city Beipiao city Gongchangling district Yangjiazhangzi Nanpiao district		
Jilin	Liaoyuan city Baishan city	Shulan city Jiutai city Dunhua city	Erdaojiang district Wangqing county	
Heilongjiang	Yichun city Great Khingan region	Qitaihe city Wudalianchi city	Hegang city Shuangyashan city	Xunke county Aihui district Jiayin county Tieli city
Jiangsu			Jiawang district	
Anhui		Huaibei city Tongling city		
Jiangxi	Pingxiang city	Jingdezhen city	Xinyu city Dayu country	
Shandong		Zaozhuang city	Xintai city Zichuan district	
He'nan	Jiaozuo city	Lingbao city	Puyang city	
Hubei	Daye city	Huangshi city Qianjiang city Zhongxiang city	Songzi city	
Hu'nan		Zixing city Lengshuijiang city Leiyang city	Lianyuan city Changning city	

(continued)

Table 1 (continued)

Province (district, city)	The first batch of 12 (2008)	The second batch of 32 (2009)	The third batch of 25 (2011)	For the forest region of Great Khingan and Lesser Khingan mountains, refer to nine cities enjoying policy privilege
Guangdong			Shaoguan city	
Guangxi		Heshan city		
Hainan			Changjiang county	
Chongqing		Wansheng district	Nanchuan district	
Sichuan		Huaying city	Luzhou city	
Guizhou		Wanshan Special district		
Yunnan	Gejiu city	Dongchuan district	Yimen county	
Shaanxi		Tongchuan city	Tongguan county	
Gansu	Baiyin	Yumen city	Honggu district	
Ningxia	Shizuishan city			

Source National Development and Reform Commission

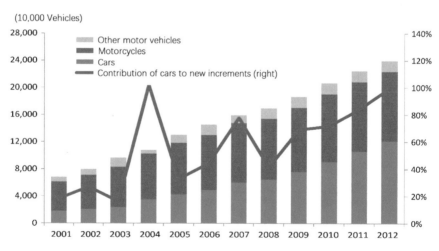

Fig. 3 Growth structure on vehicle parc in China (*source* Institute of Traffic Management Science under Ministry of Public Security)

and users of medium and small companies would opt for software services following need to realize low-cost of software and information sharing. Smart city has become a hotspot to spur urbanization development all over the country. With the profound use of modernized information technology such as testing soil for formulated fertilization, and smart monitoring of facility agriculture, the agricultural modernization level has been constantly improved.

Fourth, information consumption has gradually become a pivotal origin of point for extending domestic demands. Effect of domestic information consumption on driving forth economic growth is more prominent: from January to October in 2013, shipments of smart phones in our country reached 0.348 billion, up by 178% YoY which brings along the access of TD multimode core to commercial use; mobile internet access flow reached 1 billion G, up by 68.9%. In the first three quarters, e-business market scale reached 7 trillion Yuan, up by over 20% YoY.

Fifth, information technology has been constantly improved in terms of its immersion to the public service sector. Internet television, mobile internet learning terminal, on-line course, We-Class and other new education means and models keeps emerging; health information platforms have realized interconnectivity. Medication cloud and other technologies further boost informationization, by applying level of medication and hygiene sector. Some medical outfits with conditions are attempting to use mobile intelligent devices to boost clinical nursing level. Subsistence allowance and medical assurance have been constantly innovated, by introducing cloud computing, big data and other information technology, mode on integration, analysis and sharing of such information as endowment.

But it should be consciously awared that growth indexes of informationization development also show a downward trend that should be paid high attention to (see Fig. 4).

3. **Leap-frog development of internet industry**

In the recent years, domestic internet industry has been showing conspicuous effects (see Figs. 5 and 6). Other than enhanced strength, a large quantity of new applications and modes also show up. Though it is still lagging behind advanced countries in mode innovation, industrial application and so forth, yet it is growing increasingly close with entity economy, in particular the manufacturing industry, and has become a core power spurring development and reform of the manufacturing industry.

First, from simple copycatting to improvement, internet industry has been showing rapid development. Internet service industry in our country has realized leap-frog transformation from 'emulation of product types and simple replication of business model' to 'progressive innovation of products and original use of business model'. It has become the second largest internet power and its mainstream application is dominated by domestic companies. In 2013, Chinese companies took 8 out of Top 30 places in global market values of internet.

In September, 2014, Alibaba Group was successfully listed in New York Stock Exchange. With its market values tallying 238.332 billion dollars, it surpassed

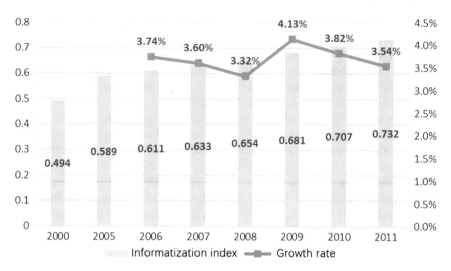

Fig. 4 Growth of informationization development index in our country (*source* State Statistics Bureau in 2012)

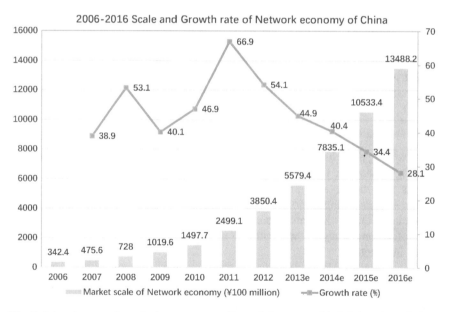

Fig. 5 Internet economic scale in our country (data of the years with "e" is estimated data) (*source* iResearch)

Facebook, Amazon, Tencent and eBay, and become the second largest internet company only to Google. The success of Alibaba can't do without constant innovation on its business model. Business model in Alibaba can be briefly summarized

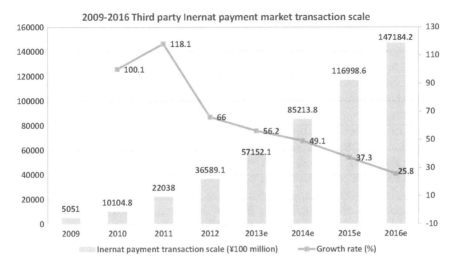

Fig. 6 Third-party internet disbursement scale in China (*source* iResearch)

as C2B2B2S, or 'Customer-Business-Business-Service'. Taobao.com adopts B2C and C2C, Alibaba adopts B2B, Alipay and Yahoo adopts B2S. Following strategic configuration in Alibaba Group, e-business ecosystem in Alibaba in the future will be a platform featuring openness, synergy and mutual prosperity. Trough opening platforms, Alibaba would appeal to various types of providers of e-business services who would bank on their professional knowledge and experience to forge individualized services catering to demands of business employees (Ali Research Center 2009).

Second, along with in-depth popularity of technology application, emerging types and service types of internet have been speedily showing up. With Microblog, WeChat, location services, mobile application shops and so forth showing up and mobile e-business, C2B2C[4] social marketing and other e-business new model being repetitively launched, internet financing services have been mushrooming. Online and offline integrated services (O2O) have been greatly arousing actual-time consumption demands and open cloud platform of internet companies is spurring new model of productive services. In 'Singles Day' in 2013, physical store of Suning Corporation surged by four folds and Tmall fared a sales volume reaching 35.1 billion Yuan, up by 83% year on year. Development platform of Tencent

---

[4]B2B2C is the abbreviation of the term "business to business to customer", a new way of network communication sales. The first B in the term refers to the generalized seller (ie., finished products, semi-finished products, and material providers, etc.); the latter B refers to the trading platforms which provide sellers and buyers platforms of contact as well as high-quality additional services; C refers to the buyers. Sellers include both companies and individuals, which reflects a logical trading relationship. B2B2C mode covers the mode of B2C and C2C, thus is more integrated and more capable of better service.

accumulatively allotted a tally of 3 billion Yuan to its developers, and Baidu open cloud services have clustered over 300,000 developers.

Third, internet industry has entered the new stage of converting from livelihood services to productivity services. By June, 2014, scale of netizens in our country reached 0.632 billion. Netizens show increasingly mature age structure and are no longer special groups. They overlap with common residents and more profoundly affect physical production and consumption process. Internet application during production and operation round of companies is intensified and the bulk of companies hold that the new generation of information technologies such as cloud computation and big data will bring about more business opportunities. Internet-based corporate globalized ecological system is renovating manufacturing industry, retail and wholesale industry, logistics and other sectors, quickening promotion of internet, intelligent, flexible and service-based transformation of traditional industry and forming new industrial organizational means and new company-user relations. Haier, Millet, Sany Heavy Industry, SPZP, Baidu, Alibaba and so forth have become leading companies putting into effect integration and innovation of the internet and industry.

2.2 Challenges Facing City Construction in China

In the 21st century, fast economic and social development leads to worsened ecology, scarce resources, financial turmoil, natural disasters and public security issues. Given the fact that domestic cities do not have self-adjustment functions, it's especially difficult to grasp complication of them during the development course. Along with development and application of new technologies, notably information technology, as well as demands for sustainable and sound social and economic development, city development will greet the following challenges in the future.

1. Challenges of speedy urbanization on city development mode

Our country is undergoing a massive urbanization. To analyze from city demographic scale and total economic volume, it is starting to transfer from a peasant country to a country with city residents, and has entered the model dominated by cities with rural development spurred by cities. Domestic urbanization features such characteristics as large scale, quick speed, multiple issues and high requirements. Since the Reform and Open-ups, urbanization rate was 17.92% in 1978, 26.41% in 1990 and 53.73% in 2013. For quite some time in the future, our country will still be going through the stage of speedy urbanization. It is forecasted that by 2015, urbanization ratio in our country will reach over 55% and this figure will shoot up to around 60% by 2020 and 70% in 2030. But speedy urbanization leads to pouring of a whopping sum of population and incurs clashes between population convergence and city education and humanistic qualities, contradiction between city development and natural resources, contradiction between indigenous culture and external culture. As a country with highest number of population, China is intertwining with globalization, marketization, informationization and so forth in its urbanization process, which brings about issues concerning with relation between urbanization and domestic economic and social development as well issues concerning with relation between urbanization of people's livelihood and production.

2. Challenges of economic transformation on adjustment of city industrial structure

Whether China can have a fast and speedy development in the upcoming two decades depends on the development of Chinese cities. In the past three decades, manufacturing industry has been the core of industrial development in Chinese cities. Manufacturing industry is still of major importance to China in the next two decades, which is a major difference from western cities. We all reckon that industrial structure of the city accounts for 3000 dollars, 5000 dollars and 10,000 dollars per capita in different stages. The development of each city has its own characteristics which need to keep abreast with the times. A key for industrial structural adjustment is adjustment of product structure, which would pose a great challenge to us. City intelligence building can be carried out via technological means and by building a collaborative system between virtuality and entity. To integrate the collaborative system among different sectors, and to boost an innovative competence in and between cities, to constantly decreasing physical increment and improving value augmenter and added value. To convert from extensiveness to intensiveness development mode, and to facilitate industrial upgrading and industrial structure adjustment and to spur an improvement of city employment and consumption.

3. Challenges of soaring city population on resources and environment

Since 1980s, rapid urban population clustering has been imposing huge pressure on regional resources and environment. Extensive growth city development model can no longer satisfies the demands in the new era. The clash on incongruity between resource environment guarantee competence building and urbanization has been increasingly conspicuous. Resources and ecological environment can't satisfy requirements on the huge pressure brought about by urbanization. And we can't continue to follow the previous old route of urbanization during the early stage of industrialization stage at the early stage of western countries. China should adapted to the new development route that contains high technological content, provides good economic profits, consumpts less natural resources, produces low environmental pollution and demonstrates edges of human resources in an all-around way. The 18th Party's Congress explicitly proposed 'vigorous promotion of urbanization and adherence to an urbanization road featuring Chinese characteristics.' Under this background, in which way shall the cities in China seek for the transformation of the development mode? What is the trend for transformation of city development model? All those add up to challenges for decision-makers and constructors of city planning, and those issues are in urgent need of their reflections and resolving.

4. **Challenges of citizen's improving livelihood level on existing public management and services**

The livelihood of urban citizens has been significantly improved as the medical hygiene condition, living condition, education and culture level, labor condition and guarantee of holidays and festivals, incomes, consumption level for clothing, food, accommodation and traveling, conditions for travelling, social guarantee level and people's freedom are all improving constantly. As development of information technology deeply aligns with people's work and life, people's urge for changing grows stronger, their demands for participate in city management are more conspicuous and social management in cities becomes more complicated. Hence requirements on city public management and public services are fairly higher than ever, thus raised the call for higher attention.

5. **Challenges of city planning concept changement on city construction**

The constructions that left over by our forebears have almost been removed. Few cities can stand the racket of the time and history, and stands to the 22nd century to become famous centuries-old cities, such as London, Paris, Chicago, Prague and Florence. Foreign scholars reckon that 'Chinese cities went through three decades of Soviet Union modernization and then went through three decades of American modernization. From the perspective of architecture, both of them might have their own defects.' For the moment, city development has witnessed an unprecedented peak, especially after the presence of the economic globalization and informationization. It provides new routes for city development form different perspectives, thus many concepts of city development emerged after that and these concepts would, to a certain degree, affect the development trajectory of cities. Ecological city stresses on the full integration of technologies and nature while hospitable city stresses on the collaborative development on economy, society, culture and environment. Intensive cities lay emphasis on routes to address living and environment issues. All those should be taken into account in city building and planning to guard against waste of natural resources, blind expansion, damage of ecological environment, same looks of city pattern, geological features and gradual vanishing of culture. All these poses severe challenges to urbanization level in China.

2.3   Basic Profile of 'Smart City' Construction in China

In August, 2008, IBM released its proposal on *Smart Earth Wins the Earth* that officially unfolded the prelude of Chinese strategies. In the proposal, IBM has been forging six solutions of intelligence bespoke to China namely: 'smart power', 'smart medication,' 'smart city', 'smart traffic', 'smart supply chain' and 'smart banking'. Since 2009, all those 'smart' plans have been constantly carried out in many Chinese cities.

National Development and Reform Commission, Department of Industry and Information Technology, Ministry of Science and Technology, Ministry of Housing and Urban-Rural Development and other ministries as well as provincial and municipal governments have been attaching high attention on building of 'smart cities'. Many cities at the municipal level and above have officially proposed building of smart cities in their planning of 'the 12th Five-Year-Plan' or their governmental work report. Over 80% of second-tier cities have explicitly proposed the development goal of 'smart city'. By September, 2013, a total of 311 Chinese cities have or plan to build smart cities with a scheduled input surpassing 2 trillion dollars. See Fig. 7 for relevant conditions.

With emergence of the concept of 'smart city', different parts in China have gone through the tide of 'smart city' and its main impetus mainly hails from the urge of local government to resolve many urgent issues brought about by urbanization. A tally of 41 cities at the municipal level and above involve building of smart cities to their '12th Five-Year-Plan' or governmental report including 10 sub-provincial cities: in the 14th Beijing International Technology Expo, 'smart Beijing' information technology made its debut. During the '12th Five-Year-Plan' and '13th Five-Year-Plan', Beijing will roundly construct 'smart Beijing' when all people will enjoy a 'smart' life; in 2011, Shanghai Municipal Government announced its 'initiative for 2011–2013 on propelling smart city building'; in 2010, *Ningbo launched Decisions on Building of Smart City by Ningbo Municipal Commission and Government* that took the initiative to systematically carry out smart city building. Moreover, Hangzhou, Guangzhou, Wuhan, Xi'an and other cities also come up with their respective thoughts of building smart cities.

Viewed from the contents, cities would reinforce basic communication network building in cities and boost broadband of communication network and its coverage; and on the other hand, it would provide some application services in key sectors such as public services, social management, traffic, power grid, medication, logistics, home furniture and so forth.

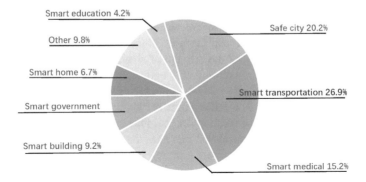

Fig. 7 IT investment structures of smart cities in China in 2013 (*source* CCID consulting, April, 2013)

2.4 Issues Confronting Construction of 'Smart Cities' in China

It needs to be noted that the major task of advanced countries is that implementation of 'smart cities' is mainly clustered in the sector of informationization due to development of information technologies and application. China is going through synchronous development of industrialization, urbanization, informationization and agricultural modernization for the current stage when interest of city managers lies on facilitating of intelligent development in cities and their understanding on building of 'smart cities' aims at the whole city. But IT companies are interested in sales of intelligent system in cities and their proposals on 'smart city' stresses technological plans and aim at 'putting into effect' their 'solutions'. It carries significant commercial nature and overlooks main bodies of the construction namely cities. It ignores complication of the huge system of cities that leads to big disparity between mayor's expectation and goals of building plan in IT companies. For example, after carefully visiting introduction of an information company on various 'smart systems', a mayor in the city made the following comments disapprovingly, 'your smart city is short of a mayor's horizon'. It can be reflected that a mayor's horizon is mainly on development decisions of the whole city rather than solutions to technological issues. By the same token, demands and expectation of functional departments of the government, companies, and residents and so on fail to align with 'smart city' building and the phenomenon of unideal effects exists.

In the present world, there's no mature developing mode for the construction of cities, thus each and every city in China shall stay reasonable, particularly confront with the fact that many relevant IT companies would carry out various publicity for their own interests. In summation, there are still many leftover problems for the construction of 'smart city' building, which includes:

1. **Lack of correct understanding, severe 'following-suit construction' and 'repetitive building'**

The concept of smart city is rooted from the commercialized behaviors of IT companies, therefore, when government wants to reuses this concept, it requires re-positioning. The government can't blindly follow suit or revolve it to a city tag behaviors, namely, whopping of the concept of smart cities but overlooking of its connotation and essence. The building of an advanced network infrastructure and the installation of a string of information technologies with high technologies can't be equal to smart city. In fact, some cities set about following suit before they actually get the connotation of smart cities, and they are plagued by vague concept, ambiguous prospect and ill-considered planning that leads to wastes and low efficiency of smart city building. Furthermore, insufficient knowledge but high expectation leads to local governments' eagerness for quick success and instant benefits, so that they expect to alter an entire city without a span of several years or even anticipates that a city with uncompleted digital city infrastructure can grow into a smart city to meet their expectation. Huge building pressure and investment risks are concealed in their planning and slogans. Internet of things, cloud computation and other industries directly related with 'smart city' have been constantly amplifying and some key cities still configure their demands of 'intelligence' and their demands for 'city building'. By September, 2013, 311 domestic cities have built or intend to build smart cities. More than 30 provinces and cities have listed internet of things as a key for industrial development and more than 80% of cities have taken internet of things as a dominant industry, which revealed a tendency of overheat. Experts have showed concerns on the phenomenon of 'repeated construction' at one swoop and reckon that overheated internet of things, cloud computation and 'smart city' might result in new surplus of output.

2. **Lack of unified deployment and mechanism innovation, severe 'information isolated island' featuring barriers between higher and lower levels**

As smart city has not been incorporated to the planning of national development, disparities of knowledge and positioning arouse as different departments and regions have varied views on the contents of smart city building. For example, the sector of city planning and building usually carries out from applying the new generation of information technology to city planning and building; department in charge of informationization would make the plan from the perspective of combining the industrialization and informationization; local governments such as municipal government would plan it from the aspect of advancing its local economy and society. At present, building of smart city in our country is independently carried out by operating departments in government. The operating department of each city builds its own smart system according to the business needs thus different sectors are still not sharing a unified construction and development management, so many public information between system can't be shared and end up with 'information isolation', which drastically brings down system building and investment profits. Furthermore, cities are now acting on their own accord and form their own systems; therefore they are short of due consistency and coordination. Different management departments and companies in the same cities are also confronted with such issues as unhindered information sharing and low efficiency of data use.

3. **Lack of core technologies and standards and formation of potential hazards on information security**

At present times, China is still deficient in terms of core technologies of relevant application and products in smart cities, for example, China still face a deep gap in terms of Telematics, LBS, GIS, family network, ITS, RS, key equipment manufacturing of chips, intelligent communication and control, massive data disposal and so forth that form core key technologies in smart city. In the meantime, compared with western countries, there is a great gap concerned with inconsistency in standards, backward standards insufficient standard building and deficient standard collaboration in the construction of smart city. For example, our country is still at the inception stage in terms of internet of things and is yet to form an industry. European countries, America and Japan that were early in advancement of internet of things have already mastered technologies of key rounds of internet of things (remote, transmission network and application computation) and core patents. For the moment, around 80% of remote core chips are bought from the occidental world and Japan. Moreover, some advanced countries force out strategic logic of Huawei and Zhongxing that make us realize all the more that 'smart city' building in our country is confronted with huge information safety security problems. Those threats are not just from those internet communication companies that take a monopoly control in the sector of hardware amenities in China, but also from their dominant role in information business system, database management and business solutions and our shortage of effective monitoring and solving ways as regards their reliability and safety issues. Experts generally hold that existing information safety protection system in our country has huge loopholes, which actually formed threats to the informational safety of the country and citizen.

4. **Lack of technology and management system to tackle with big data challenges**

Data are increasingly becoming the foundation for social wealth and innovative development, while big data becomes an important impetus for upgrading of existing industries and birth of new industry. In an intelligent era spurred by data, big data constitute major means of production. Wise utilization of data can help us to obtain a breakthrough in intelligent production and living competence. Big data has also triggered a re-evaluation on scientific research methodologies in the science and technology circle that is incurring a revolution of scientific research ideology and methods. Mass data in the internet of things should be deemed as a country's strategic resources. Experts believe, in the internet, who can validly monopolize data resources, and then who can rein in the world. In technologies and management system, existing technologies on the data center stop short of satisfying demands for big data while growth of storing competence can't match expansion of data. Governmental outfits, industrial organization and large-scaled companies are yet to establish special data governance outfits to coordinate data governance. They can't attain open data or make sure that the most substantial production materials in a big-data era to freely flow around, so as to incur innovation and propel development of knowledge economy and online economy, and to facilitate conversion from extensiveness to intensiveness.

5. **Lack of in-depth fusion of technology and business, management and services, and failure to incorporate residents' appeal to the building agenda**

In the incipient stage, both academic research and governmental and commercial practical behaviors put emphasis on information technology building. They intend to improve city economy and management efficiency via information technology. It can be said that the concept of smart city during this stage is mostly guided by technology, which belongs to the range of digital city and stressed hard on the strength of city building. The aim for smart city shall be providing hospitable and handy living environment to residents, thus blindly goes after technological advancement and overlooks application of technologies turn out to be one of the errors for smart city building. How to improve overall social functions of cities via intelligent technology application, facilitate overall social functions of the cities, boost overall social functions in cities, improve a city's humanistic quality and cultural foundation to publicize cities to provide a more livable life environment to residents, shall be the prime tasks for soft power building in a city. Hard power building and soft power building together helps to forge the sustainable competitiveness for a city.

3 Several in-Depth Analyses on Intelligence Development in Cities

The aforementioned conditions lead us to a clear awareness that social and economic development in our country is subject to various objective conditions, so it requires that our country shall no longer repeat the previous development route and call for new breakthroughs in the new setting. Constrained by the city's own development conditions, changes of city development conditions call for breaking of restrictions on routes of city development without ridding the city's featured blind development in order to realize new development objectives; globalized development incurs more and more global issues, global risks, global financial crises, global climate warming, environmental pollution, information safety and other non-conventional safety threats. It required us to formulate various countermeasures to intelligent development in cities in the globalization setting. With city development in the future, it should pay high attention to maintenance and upkeep of cities, realize 'co-emphasis of building and use' and bring benefits to people living in cities. In summation, forces spurring city development in the future will witness great changes and a variety of brand new city forms will take form. Based on causes of 'smart city' statue of China and abroad leads to the following conclusion:

3.1 Essence of City Intelligence Requires Collaborative Development of 'PHC'

City intelligence reflects a new orientation of city development under the new technological revolution condition. It should surpass pure viewpoint on information technology and carry out city building from a broader vision. Studies indicate that city intelligence development should be taken as a complicated system via coupling and correlation of three dimensions. The first dimension is physical space, i.e., a physical environment and city substances in a city; the second dimension is 'human society (H), i.e., human decision-making and social exchange space and the third dimension is 'cyber space' or the 'virtual' space comprising computer sand internet. City intelligence can be interpreted as a process of PHC synchronously, and mutually spurring one another.

Traditionally, the city building lays its emphasis on the first dimension, such as city planning, building, management and so forth. Its major emphases lie on physical square and facilities in a city, land use, layout of functional regions, transport planning, energy, environment, water resources, city infrastructure, etc. Digital city, network city and other building proposed in the 1990s all commenced to focus on building and extending of the third dimension. Building of intelligent cities carried out in different countries and domestic cities is, in actuality, concentrated on the forging of the third dimension.

It should be noted that city building revolving on the first and third dimensions are, without doubt, important, but consolidation and expansion of the third dimension as well as interconnectivity of the second dimension with the first and third dimensions will become a more profound and far-reaching topic for city intelligence. City intelligence building in the future should pay attention on second dimensions building while it stresses first and third dimensions building. It should also punctuate coupling among three dimensions and position the center of city building as round development of mankind, notably comprehensive improvement of man's innovativeness before spurring comprehensive and sustainable economic, social and ecological development.

It can be seen that the current city has entered PHC from PH, which is a micro tendency for city intelligence. Only the name, contents and development phase are different in different countries. Construction of 'smart city' both in China and overseas is mainly clustered in forging of a third dimension while domestic city intelligence is mutual collaboration of 'three dimensions' to facilitate harmonious development of the society, economy and ecology and to surpass the building concept on digital city, network city and smart city.'

3.2 Advent of a New Era Marked by New Technological Revolution's Facilitation of Intelligent City

There are plenty of viewpoints on new technological revolution in the current world, includes:

Brian Arthur, an American economist, proposed the concept of 'second economy'. Its main contents cover processor, chain connector, sensor, actuator and economic activities operating on it that form the second economy (not virtual economy) other than physical economy (first economy) people are familiar with. The essence of second economy is to attach a 'nervous layer' to the first economy so that national economic activities can gain intelligence. It marks the biggest changes of electrification since a century ago. He also estimated the scale of second economy and he reckoned that by 2030, the scale of the second economy can approach the first economy. The values of information technology is not restrained to traditional hardware, software and services and that information technology gets incorporated to people's society and the physical world features far more value space. To attain these values, Google and other companies adopt socialized big production model with low-cost information services that are not seen in industrial capitalism, i.e., it would enable thousands of millions of customers to work for it free of charge so that customer activities serve as their production activities. —From academician Li Guojie, July, 21, 2013, academic report in Guiyang entitled *intelligent city construction and Big Data Technologies* (Li 2013).

The 'Third Industrial Revolution' concept is mainly proposed by American Industry Circle and the academia. Human society has gone through the stream engine and electrification ear to the emergence of 'Third Industrial Revolution' with massive application of information technology. Its main viewpoints can be summed up to four types: first, according to Professor Vivek Wadhwa from Singularity University, combination of artificial intelligence robots and digital manufacturing technologies will incur a revolution on the manufacturing industry; second, according to Economists, the magazine, the world is going through the 'Third Industrial Revolution' with its core marked by gradual popularity of digitalized manufacturing, new software, new techniques, robots and internet services and myriads individualized, scattered and adjacent production being major characteristics. It means that the traditional production means with large scale assembly line will be terminated; third, according to Jeremy Rifkin: internet technology and renewable energy combined will bring about the 'Third Industrial Revolution' and hence bring major transformation of people's production life and social economy; fourth, Report on Prospects of 2030: How America Tackles with the Strategy of Technological Revolution in the Future co-compiled by Burrows and Robert Manning, it reckons that the world is now standing at the junction of major technological revolution. Thus the 'Third Industrial revolution' with manufacturing technology, new resource and smart city as representative will exert major impacts on shaping politics, economy and social development.

The industrial circle and the academia in Germany proposed the concept of the 'Fourth Industrial Revolution'. *Seizing of Future of Manufacturing Industry in Germany—Suggestions to Implement 'Industry 4.0' Strategies* was co-launched by the industrial circle and academia circle (German Academy of Engineering, Fraunhofer Society, Siemens and so on) in Hannover Messe Expo in April, 2013. 'Industry 4.0' project was one of the ten major future projects determined by *High Technology Strategy 2010* by

German government in July, 2010 that intended to back up research, development and innovation of the new generation of revolutionary technology. Its main contents cover that the first three industrial revolutions hail from mechanization, power and information technology. Applying of internet of things and services to the manufacturing industry is triggering the Fourth Industrial Revolution. In the future, companies will establish a global network that fuses their robots, memory system and production facilities to CPS, a virtual network. In the manufacturing system, the CPS or virtual network incorporates intelligent machine, memory system and production facilities that can independently and automatically exchange information, trigger actions and controlling. All those facilitate fundamental improvement of manufacturing, project, material use, supply chain and management at the life cycle. —From *Securing the future of German manufacturing industry Recommendations for implementing the strategic initiative INDUSTRIE 4.0* by Final report of the Industrie 4.0 Working Group (Kagermann et al. 2013).

The team leader of Qian Xuesen proposed the concept of the 'Fifth Industrial Revolution'. The first revolution: it occurred in the middle 18th century that was dominated by conversion from thermal energy to mechanical energy via steam engine. Britain took the initiatives to transit from agricultural economy to the industrial economy and hence seized the first place in the world in total industrial output value; the second revolution: from the end of the 18th century to early 19th century. It was dominated by steel technology and mechanical manufacturing technology. Germany can match UK; the third revolution: from middle 19th century to the late 19th century. Its dominated technology was electrification. America grew to the most representative advanced country in the world. The fourth revolution: it occurred in the 20th century. It was dominated by semiconductor and laser. Japan grew to the fastest-speed economy giant after the Second World War; the fifth revolution: it occurred from the middle 20th century to the early 21st century. What is the dominant technology? In *Scientific American* for September, 1991, a special magazine on *Communication, Computer and Internet* was published that analyzed development tendency of information technology in the 21st century and its epigram was work, learning and booming development in a cyberspace. After times of wartime practice and researches, the Obama administrative decided to put Cyberspace strategy as the highest preference of the country and defined 2009–2010 as a strategic leap-frog development year for Cyber on May, 29, 2009. It then commenced to plan out Cyber strategy for 2020–2025. The core factor of Cyber space was not just to enhance the function of physical level and transmission level of information, rather, it should pay more attention on the cognition and decision-making level of information and give play to the effect of man in Cyberspace. It aligns with the core factor of 'information environment with man-machine congruity.—From *The Fifth Industrial Revolution and Cyber Strategies* in the academic report by academician Wang Chengwei on July, 29, 2013 (Wang 2013).

Relevant academicians in the Chinese Academy of Engineering reckon that such viewpoints as 'second economy', 'third industrial revolution', 'industry 4.0' and 'fifth industrial revolution (cyber strategy) and so on vary in statement, yet their expectation on supporting role displayed by information technologies is basically consistent (or gradually converging). In other words, it should not just improve functions of the physical level and transmission level of information technologies, it should enhance functions of the cognition and decision-making level of information technologies; it should stress building of 'man-machine congruent environment' dominated by man and foster and control qualifying talents that can fit the environment. On this ground, development tendency on relevant technologies will come to this conclusion: the key of big data is on how to speedily extract effective 'strategies' from massive 'data' based on people's demands; the key of internet of things is not to connect 'things' but to ensure realization 'man-machine (material) congruity' with 'consistent time and space' and 'fusion of isomerism information'; demands for information safety and credibility will witness substantial changes and cloud computing will become a sector that should draw most attention from safety and credibility. The key of state-level information infrastructure is 'independence, controllability', 'self-adaption' and 'self-renovation'.

In actuality, one of the major characteristics sis to fuse the new generation of remoter technology, internet technologies, big data technology and engineering technologies to various systems of the city before forming city building, city economy, city management and upgraded development of services before greeting intelligence, a new era for city development.

According to Stiglitz, a Novel-prize winner, in the early stage of the 21st century, two major events affecting the world cover new technological revolution of the US and urbanization in China. 'Urbanization in China and new technological revolution will not just facilitate advent of intelligent era in Chinese cities and might bring about birth of more technologies in China. China will, without doubt, greatly contribute to economic and technological development of the world.

Fig. 8 Comparison on development trajectory between China and advanced country

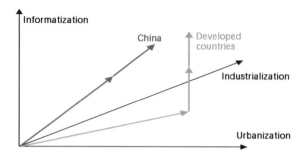

3.3 Use of 'Intelligent City' to Replace 'Smart City' Is More Fitting to Chinese National Condition

IBM proposes the concept of 'smart city'. Smart means smartness and cleverness that does not amount to wisdom. Secondly, the western world has gone through large-scaled urbanization and industrialization era and does not ask for larges-scaled infrastructure building. Major task of the current cities lies on intelligence of management services, so administrative functions of city managers is more constrained than Chinese mayors. For the moment, our country is going through the stage marked by synchronous development of industrialization, urbanization, informationization and agricultural modernization (four modernizations) and the perplexities and issues it confronts feature uniqueness in terms of quality and quantity, so intelligent development route in Chinese cities vary from the western world (see Fig. 8). To interpret intelligent city from their perspective stops short of redressing issues confronting development of Chinese cities. It denotes that connotation and practice of city development in China is more diversified than 'intelligent city' in the west.

In actuality, the concept of 'smart city' mainly intends to apply IT system to city management process such as intelligent medical system and other concrete projects. Measured from the objective rules of city development in China, shortage of 'mayor's horizon' when it comes to building of 'smart city' would leads to failure to redress main clashes in cities. If upgrading of economic development and city planning featuring Chinese characteristics is yet to be realized, then intelligent development of cities is devoid of souls. Just conjure this picture up, if a city repetitively utilizes instruments and is short of long-term planning, how is the city built? How is the society congruent? How does economy rise? And what is defined as felicity for people?

Studies reckon that the word 'smart' is not fitting to development featuring Chinese characteristics and suggest that re-definition is needed by proposing the concept of 'intelligent city' or 'iCity'. During exchange and dialogue with national ministries, commissions and local government as well as experts and scholars attending topics, officers, scholars and representatives of all circles have more extensive vision and prospects on the understanding of 'smart city'. The Chinese definition of 'smart city' contrasts to the concept of 'smart city'. What they intend to do is IT intelligent system in cities. China needs to build an intelligent city, so it's suggested that China use the concept of iCity that is more fitting to Chinese national condition. For metropolis cities with rural areas, the essence of building an intelligent city is to guarantee that city grab fast, speedy and saving intelligent development that calls for upgraded in-depth fusion of new urbanization, in-depth informationization and industrial upgrade so that a city can make progress extensively in a green, hospitable, secured and sustainable way.

References

Ali Research Center. 2009. *The Coming of New Commercial Civilization—Research Report on the Network Commerce Development in 2009*. Hangzhou: Ali Research Center.

Atlantic Council. 2013. *Envisioning 2030: US Strategy for the Coming Technology Revolution.* Washington, D.C.: Brent Scowcroft Center on International Security.

European Commission. 2005. i2010: The Next Five Years in Information Society. Brussels: eEurope Advisory Group.

Kagermann, H., W. Wahlster, J. Helbig. 2013. Securing the Future of German Manufacturing Industry: Recommendations for Implementing the Strategic Initiative INDUSTRIE 4.0, Final Report of the Industrie 4.0 Working Group. Berlin: Forschungsunion im Stifterverband für die DeutscheWirtschaft e.V.

Li, G.J. 2013. Intelligent City Construction and Big-data Technology. Guiyang: Eco Forum Global Annual Conference Guiyang.

Wang, C.W. 2013. *The Fifth Industrial Revolution and Cyber Strategy.* Beijing: Chinese Academy of Engineering.

Chapter 2
Connotation and Significance of Intelligent City Construction

1 Definition of Intelligent City

'Smart city' proposed by IBM is defined as to deploy information and communication technology to monitor, analyze and integrate various key information for city operation system before making smart correspondence to varying demands covering people's livelihood, environmental conservation, public safety, city services and industrial and commercial activities. The essence of definition proposed by IBM is to utilize advanced information technology to attain smart management and operation before creating a rosier life to people and facilitating harmonious and sustainable growth in cities.

Through studies on codes of communication, Telecommunication Research Institute under the Ministry of Industry and Information Technologies defines 'smart city' as to integrate existing resources including smart integration, applied integration and remote network integration of data. Data integration breaks through information isolation, realizes information sharing at the municipal level, reinforces unified management of data, realizes accuracy and timeliness of data, establishes a system with data being converted to values and brings to fruition integration operation in cities. Based on portal of applied integration, it provides unified smart application services and realizes high-efficient synergy of operating industrial chain in the whole smart; multiple remote networks as video monitoring, sensor, RFID and so on of the network help realize unified monitoring and management on the city remote network and carry out unified analysis and optimization of operation data of city operation on this ground before realizing smart management of city operation and providing more efficient city services.

Professional commission on digital city under Chinese Society for Urban Studies reckons that it is premised on live-action model in cities, puts city buildings as the carrying subjects, integrates city resources, environment, society, economy and information to basic factors of people, companies and city facilities in cities and

© Springer Nature Singapore Pte Ltd. and Zhejiang University Press, Hangzhou 2018
Y. Pan, *Strategic Research on Construction and Promotion of China's Intelligent Cities*,
Strategic Research on Construction and Promotion of China's Intelligent Cities,
https://doi.org/10.1007/978-981-10-6310-7_2

utilizes internet of things to acquire dynamic city operating data. It aggregates various industrial applications in the public information platform of smart cities.

Global Trend 2030 defines 'smart city' as that it is a city ambiance that makes use of advanced information technologies to realize maximum city economic efficiency and most promising life quality with minimum resources and at the cost of environment degradation. It can be taken that this definition highly sums up knowledge on generality of smart city against different backdrops of information technology, industrial economy and systems.

At present, understanding on 'smart city' varies that can be generally categorized into three viewpoints namely projects, in-depth informationization and city system.

iCity proposed by our study starts from 'CPH'. With integration of various data, on the ground that digitalization, networking and intelligence and through high integration and in-depth integration of knowledge technology and information technologies, it carries out effective services following economic and social development and demands of residents that form inexhaustible impetus for finding and resolving issues, adding more vibrancy and sustainable trait to cities and form new ideology and model of city development. It not just gives direct interests to residents from the perspective of economy, society and services. It can bring them to feel convenience, synergic efficiency, congruent and sound greenness and visible safety. Social values of intelligent city are mainly reflected in that they can effectively resolve city diseases, extend industrial development sectors and bring more satisfaction to people's employment and entrepreneurship. The economic values of intelligent city are mainly reflected in that it is a multiplier of city economic surge.

> iCity we have proposed is defined as to scientifically plan and arrange city CPH, deftly cluster wisdom of city residents, companies and administration, deepen mobilization of comprehensive city resources, optimally advance city economy, building and management, continuously improve city development level and people's livelihood and better serve the purpose of the now and future of city residents. In short, sound use of CPH helps boost city development level and people's livelihood.

2 Main Characteristics of Intelligent City

Intelligent city is a new pattern for city development that puts internet, internet of things, telecommunication network, power grid, wireless broadband and other network combination as the basis, takes high integration of information technology and comprehensive use of information resources as the major characteristics and intelligent technologies, industries, services, management and life as important contents that is dedicated to self-amendment and timely addressing of key issues in city economy, society and ecology. Its main characteristics cover.

People oriented: It takes people's demands as the basic starting point, promoting social progress by individuals, put people's development as the basis, realizes digitalization and intelligence facing the future and enables people living in cities handier and more secured.

Comprehensive perception: It utilizes comprehensive intelligent sensing to conduct actual-time monitoring on the city's core systems and to carry out comprehensive perceptions on the physical cities and attains intelligent acquiring of all information in the physical city.

Interconnection and interworking: Internet of things helps realize interconnectivity of all information in the city.

In-depth integration: with integration and connection of internet of things and the internet, multi-source heterogeneous data can be fused into consistent data.

Collaborative operation: while constantly consolidating and working on city infrastructure, it fully utilizes city intelligent information system facilities to realize high-efficient and collaborative operation of CPH and to ensure normal operation and sustainable and sound development of cities.

Intelligent services: extensive, actual-time and intelligent services. On ground of intelligent information utilities, it utilizes big data and cloud service new model, fully utilizes and mobilizes all information resources and disposes mass perceptual data, dig into data and tackle with knowledge by forming a new service model and a new structure of service system. It provides various levels of low-cost and high-efficient intelligent services or decision-making and cognition services to people (mainly the government, companies, residents and so on).

3 Important Significance of Construction and Propelling Intelligent City in China

When perceiving development history of economy in advanced countries, they confront different development opportunities in different development stages. Economic advancement in its true sense in our country started in 1978. During the period, we have been grasping many development opportunities and grabbed sound development. For the moment, we are going through the pivotal period for economic development. We can't possibly grab the development opportunities advanced countries used to confront, yet we should be aware that different stages have different development period. Our country is going through the pivotal stage for economic and social development, which are also development opportunities. Opportunities can facilitate synchronous development of urbanization, industrialization and informationization, becomes major foundation to realize new regional development, can propel industrial transformative development; can redress actual and urgent world difficulties. We hold that building and promotion of intelligent city defined by China might redress the difficulties and have practical and far-reaching historic significance.

3.1 It Can Propel Synchronous and Coordinated Development of Urbanization, Industrialization, Informationization and Agricultural Modernization

Report of the 18th Party's Congress indicated that we should 'adhere to taking a new modernization route on industrialization, informationization, urbanization and agricultural modernization featuring Chinese characteristics, propel in-depth fusion of informationization and industrialization, benign interaction of industrialization and urbanization and mutual coordination of urbanization and agricultural modernization, and facilitate synchronous development of industrialization, informationization, urbanization and agricultural modernization.' The epochal characteristic for the moment is synchronous development of 'four modernization'. iCity is at the junction of 'four modernization', so it has become a major development opportunity in our country. To facilitate iCity building and better display advantages of city management system and mechanism displays a positive and auxiliary effect to roundly deepen system reform and development that can be taken as a basic platform for synchronous development of 'four modernization', become a major seizure for economic and social development in our country, skip the so-called 'middle-income trap' (Fig. 1), takes new urbanization development route featuring Chinese characteristics and imposes major impact on the world.

3.2 It Can Become a Major Foundation to Realize New Regional Development

Regional economic development strategies feature such characteristics as entirety, comprehensiveness, systematicalness and dynamic nature. The development goal of regional economy is to establish a criss-cross and mutually dependent internet system with distinct hierarchy. After the 18th Party's Congress, the Party and the Country considers the situation and proposes the strategic ideology on development of new regions such as Silk Road Economic Belt, Maritime Silk Road, China (Shanghai) Free Trade Pilot Zone, Beijing-Tianjin-Hebei synergic development, Yangtze River Economic Delta and so forth. From spot to line to plane, covering the land ocean to the overseas and ranging from the seaboard to the inland to the frontier area, these national strategies put the concept of 'domestic and overseas linkage, inter-regional collaboration, co-existence of external synergy and internal synergy' as the commanding force, breaks pure limitation of administrative areas or

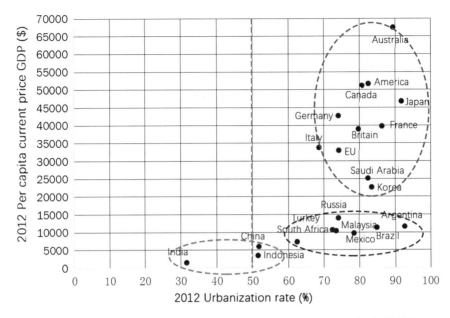

Fig. 1 Condition on urbanization ratio and per capita GDP in relevant countries in 2012 (*source* The World Bank)

the national border, expands regional economy planning to cross-city, cross-province or even cross-country and attempts to get rid of the limitation of administrative regions. All those will guide regional development to step into a new era. As a 'point', Intelligent city construction is a major basis to realize new regional 'line and planar' development.

3.3 It Can Facilitate Transformative Development in Cities

Technological innovation is a major means for industrial development in cities. With updated development of such technologies as information technology, automation and manufacturing technology, resource technology, health technology and so forth, city building has new methods and building methods. Building and facilitation of intelligent city is conducive to that China takes a new industrialization route featuring high scientific and technological content, good economic profits, low consumption of resources, low environmental pollution and full displaying of human resources edges. With adjustment of industrial structure and expansion of new workforce, it will get updated in qualification and knowledge which would contribute to entrepreneurship and employment of city residents and development of companies. Driven by technological innovation, city industrial development has three points of advantages. Firstly, it contributes to reducing consumption of resources, boosting efficiency and bringing down costs. With in-depth fusion of information technology and industrial development as well as research, development and application of its own advanced technologies, industrial upgrading and transformation can be attained; secondly, emerging industries in cities might emerge in the following 12 sectors namely: 1. Information sector: chips, internet, intelligent technology; polymerization technology: information and biology, cognition, nanometer and other technologies; 2. Bio-medication sector: vaccine, medicine, diagnosis technology and equipment; 3. Material sector: energy material, information material, high-performance structural materials, biological material, front-edge materials (carbon nano tube, graphene, superconduction); 4. Energy sector: efficient and clean use of various resources, energy-saving smart power grid; 5. Aviation sector; 6. Spaceflight area; 7. Agricultural sector: molecular breeding, biological rector; 8. Ocean world: maritime survey, carrying, biological resources, ecology and environment; 9. Environmental conservation sector: protection and monitoring of water, soil and gasp, recycled use of resources; 10. Manufacturing sector: smart manufacturing, sensor, intelligent diagnosis; 11. New energy car sector: battery, dynamo, power control and car structure; 12. Modern service industry. Thirdly, brand building. Brand building is not accomplished in one day. It needs long-term accumulation and sustainable marketing test. It is attained by three means namely by depending on the mass such as CNPC, CNOOC, China mobile, ICBC and so forth, by depending on long-term reputation such as Siemens, IBM, Toyota and so on, and by depending on innovation such as Microsoft, IPhone, Huawei and so on.

Case 1. The manufacturing technology marches from mechanization, automation and digitalization towards intelligence and intelligent manufacturing is a new model fusing informationization and industrialization. It is noteworthy to mention that US is going through the emerging tide of Makers, i.e., individuals use their computers to be engaged in innovative manufacturing activities via DIY and 3D printing or in other words, it is to use informationized sharing innovative (innovation and manufacturing) platform to start up their own business and regain glory. It is the so-called 'third industrial revolution' that brings the manufacturing industry back to cities. For example, a speedy molding company that works with American NASA using selective laser melting techniques to print out rocket ejector model by using nichrome alloy power to print products by layers. It reduces down from the original 115 components to only two parts and it takes less than one month from the original half a year. Cost is cut in half.

Information service industry will become a major growing pole for city industries. Massive internet application, high-speed rotation of oceanic data actual-time information interaction of mass terminals in the internet and actual-time exchange and transaction of thousands of millions of people lead the information internet technologies to March towards the tendency featuring 'broadband, extensiveness, mobility, fusion, safety and greenness', also known as 'second economy'. Super-high internet, extensive internet technology, 4G wireless technology, fusion internet technology, green internet technology, mobile internet, local area network, car networking and other new technologies, new businesses and new types based on the internet are popping up. It should arouse attention that in 2013, domestic online economic market rose by 50.9% in market scale. Internet, notably social networking, e-business and mobile internet, has brought human society to a new era with big data and knowledge mining. Data has been fused in all industries and functional areas that have become a major production factor. Big data and knowledge mining has gradually become a major constitute of infrastructure in modern society. Big-scaled data and it's processing makes possible the unimagined services and businesses and incurs access to the new market beyond imagination. China is home to a gigantic base of people and application market that characterizes high complication and is bristled with changes. China is the most complicated big-data country in the world. We need to acknowledge rules and characteristics of the big data era and probe into solutions based on big data. It is a major means to boost a city's municipal services and development efficiency of industries. It is also the best opportunity for mayors to facilitate big data and internet of things.

> Case 2. It probes into knowledge in big data and serves decision-making and coordinated development of mayors: administrative information (e-administration, community), company information (online design, network manufacturing, healthcare information), culture, scientific research, decision-making and so on; Case 3: material genome shoring up scientific research. Research and development of a new material takes 18 years from laboratory to market. But 'material genome project' is a database that establishes structural characteristics of materials. It explores the most probable compound for experimentation following property of materials needed. It can largely shorten development time of products and hunts for science rule. It can be thus seen that information service industry is becoming new impetus for speedy growth of company investment and people's consumption.

In summation, building and promoting of intelligent city can make full use of information and other technologies, which facilitates transformation of manufacturing and service industry in the new industrialization road, realization of conversion from city industries to intensiveness, slows down increment of things, quicken growth of values and improve added value (see Fig. 2). It contributes to use and integrate of various e-business, big data, cloud computation and internet of things, realizes development of internet technologies that characterizes 'broadband, extensiveness, mobility, fusion, safety and greenness', facilitates boosted efficiency of city industries, forms new production factors, helps improve qualities and knowledge of residents and opens new conditions for entrepreneurship and employment.

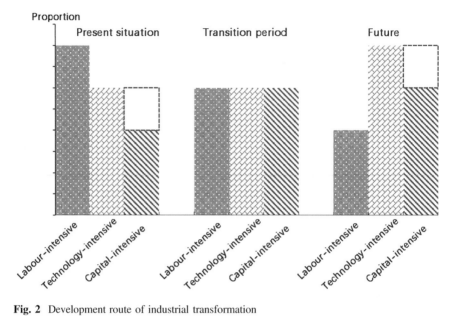

Fig. 2 Development route of industrial transformation

3.4 It Can Improve Connotation and Quality of City Management Services

The value of building and facilitating of intelligent city lies on probable use of minimum resource and time consumption to help people acquire bigger satisfaction and joy. It helps residents enjoy more considerable services, diversified and handy green and intelligent traffic, continuously integrated and efficient medical health services with controllable costs, safe and green organic food, fresh air and tidy accommodation environment, lifelong individualized interactive education, timely-seen knowledge resources and just and fair living environment without the need of stepping out of the doorway.

Construction of intelligent city helps optimize city management. The model of 'internet of things + internet + cloud computation + big data' helps timely and accurately responds to city services, management and social public security. Construction and facilitation of intelligent city helps spur big data collection and categorization of cities, form big data platform and knowledge center, advances from simple decision-making within limited information to the most optimal decisions in the city system information, helps municipal government boost city management service level and facilitates deepening of reform and development of city administration system.

3.5 It Can Raise Connotation of Efficiency, Characteristics and Culture in the City

City construction in our country is plagued with many ongoing challenges. Intelligent city construction and facilitating helps us utilize new technologies to work on city buildings, roads, environment and so forth. On the other hand, city characteristics and cultures form the glamor of a city, are city landscape and vitality of images and are seen as a unique city. Many domestic cities have attracted worldwide attention due to their distinct historic significance and profound cultural deposits. Many cities in the world are famous all over the world due to featured city characteristics and ample culture. Featured culture in cities has become a major power for many big cities to spur local economy, social balance and harmonious development. Through Intelligent city construction and facilitation, new technologies can be fully utilized to maintain a city's historic pattern, protect cultural heritage, pass through local culture and so forth and further upgrade city characteristics and culture. At the same time, it can prevent the so-called 'image projects' without local characteristics and culture such as 'city building via enclosure', 'city building in mountain', 'cement woods', 'ancient European city' and such characteristics and cultures that the local city is devoid of. Take the Dujiangyan Irrigation System as an example. Dujiangyan is the world's most ancient, largest-scaled and most efficient ecological and environmental water conservancy project that has been displaying a major role for over 2000 years. When building Dujiangyan, Li Bing tailored to the local condition. Therefore, to counter against globalized homogeneity tendency, building of Chinese city has to expend time, money and efforts to preserve its glamor.

4 Development Visions

Intelligence city construction will promote the deep development of the urban economic and social development, visions includes:

4.1 Productivity Factors Are Largely Released and Economic Development Means Are Conspicuously Transformed

Due to intelligence of production technologies, productivity has been greatly enhanced that is incurring intensive distribution of production raw materials, bringing about transformation of production means and witnessing significant transformation of economic development means. All those make people's life more convenient, bring down price of commodities needed and boost satisfaction of city residents.

4.2 City Space Layout Is Dispersed and Communication with Residents and City Supervision Goes to Zero Distance

Thanks to development of city intelligence, city space layout will witness constant dispersion and the same can be said as regards people's accommodation, but exchange among people should be zero distance. City management, service and supervision brings services provided by the Central Government, ministries and local government to companies and residents to be zero-distant and so is resident's distance with governmental supervision.

4.3 Home-Based Office Is Gradually Populous that Would Validly Diffuse Traffic, Energy Conservation and Environmental Protection

Thanks to intelligence of communication and network, a great portion of on-duty staffs can work at home. As long as corresponding planning is made, mutual win of home-based work and life can be attained. Taking white-collar staffs as an example, at least over 1.7 million staffs work at home if the ratio is calculated at 10%. It reduces congestion at the peak hours in work days (when one car for 1.7 people is calculated, a total of at least 1 million vehicles can be decreased during workdays), largely brings down energy consumption of office towers and exhaust gas emission and greatly improves work efficiency. Along with rising number of home-based office staffs and completion of city intelligent traffic system, people's peak of travelling during daytime has been converted from going to and off from work to shopping, travelling, visiting of friends and so on and enters the next new stage.

4.4 In-Depth Fusion of People's Life, Work and Study, Constant Improvement of Game Rule of People and the Society

Thanks to intelligence of information sharing technology and improvement of hinting function of social security (artificiality and nature), people are equipped with more security. Development of intelligent medication evolves from treatment to prevention and hence people's living quality is upgraded; popularity of online education leads people to anywhere they want from classroom and facilitates people to consciously accept lifetime education; development of e-business and online intelligent services help transform people's consumption, employment and entrepreneurship and lead them to pay more attention on social credit building. City intelligent make the society fairer, help profoundly fuse people's life, work and study, further upgrade and diversity people's value concept, life concept and world's concept. In terms of constrained consumption, boosted efficiency, improved environment and dredged traffic, people can pay more attention to their due social responsibilities. In terms of social order, people expect to see completion of corresponding laws and regulations.

4.5 'Cognition, Development and Use of Man's Brain' Initiatives Become a New Highlight

With the advent of digitalization and an intelligent era, new materials, techniques and technologies popping up and breakthrough of medical research, advanced countries begin to launch research and application of people's mystical brain (Obama mentioned that in the future, America might input billions on man's brain research project; similar plan might be carried out in the west). During the city intelligence process, study and gradual knowledge on people's brain as well as the behaviors of development and application together will witness a new peak that makes construction of intelligent city all the more individualized that will become a glitzy scene line.

Chapter 3
Objective and Contents of Intelligent City Construction in China

1 Guiding Ideology, Basic Principles and Idea of Intelligent City Construction and Facilitation

Guiding ideology: it takes scientific outlook on development as the guidance, sticks to the people-oriented and advancing with the times, adopts pilot first and fans out from point to area; takes a foothold on the long-term basis and century-long plan; hews to Chinese characteristics and adjusts measures to local conditions; it is dominated by the government and attended by city residents. It follows environmental friendliness, safety and health. With the guiding ideology marked by self-adjustment and handy development, it revolves upon the overall goal and requirements of building a well-off society, takes comprehensive improvement of quality and level of urbanization as its tenet, plans material, information and mental resources of city development as a whole, enhances new impetus driven by innovation, arouses vitality of market objects, establishes a new system for modern industry, reinforces guarantee competence of information safety, validly improves social management and public service level in cities, raises a city's use efficiency and comprehensive bearing capacity of land, space, energy and other resources, improves ecological environmental quality in cities, improves livelihood and felicity of city residents and soundly and orderly remotes the construction of intelligent city with Chinese characteristics.

Basic principle: top-notched design and differentiated development; coordination and implementation by stages; dynamic adjustment and combination of the reality and virtuality, equal emphasize on construction and use and focus on practical results; open operation and facilitation by effects. Intelligent city construction in China is a complicated and huge systematic project that is related with strategic issue concerning recent and long-term development. As different cities vary in development history and phase, environment and culture also varies, so we need to follow objective law, take into account near-term, middle-term and long-term development, conduct strategic top-notch designing by categories, ensure the

© Springer Nature Singapore Pte Ltd. and Zhejiang University Press, Hangzhou 2018
Y. Pan, *Strategic Research on Construction and Promotion of China's Intelligent Cities*,
Strategic Research on Construction and Promotion of China's Intelligent Cities,
https://doi.org/10.1007/978-981-10-6310-7_3

strategic, forward-looking and sustainable development of intelligent city, do not seek a comprehensive and unified development model, draw out differentiated development strategies following difference on city development, maintain cultural and environmental characteristics of cities, considers development demand of all aspects and conducts corresponding project implementation by steps following requirements of priority.

Basic idea: it is to construct a harmonious, hospitable, intensive, innovative, fair, efficient and safe intelligent city. It fully utilizes results of modern science and technology, adopts mayor's vision, cracks down the institutional barriers to city management, shatters clashes on city development, environment resource, space and so forth by smashing the limitation of 'management wall' of city piece management, breaks the bottleneck of information and knowledge acquisition for city development, constructs a completed system of organization guarantee and policy, guarantees social fairness and just, roots the new system for a city's sound and efficient development, attains coordinated development of a city's intelligent management and services, intelligent production and operation and intelligent life and guarantee, further improves comprehensive intelligence level of city infrastructures, guarantees information safety of core sectors and information system and eventually realizes satisfaction by the government and the masses and adds more extensive international influence and competitiveness to Chinese cities by advanced and safe information technology measures.

2 Objectives of Construction and Facilitating of Intelligent City

As can be indicated by Fig. 1, advanced countries vary from countries high urbanization ration and low per capita GDP. Figure 2 shows average development trajectories of countries in the first set ① and those in the second set ② in Fig. 1. It can be seen that the two share the same trajectory before urbanization rate reaches 55%. The two show differences in development rate when urbanization rate reaches 60–70%. After the climax of urbanization (at 70%), per capita GDP surges like a right angle in the former frame while that in the later frame slowly advances. Studies reckon that development in China is going through a pivotal stage. In next 15 years, whether industrialization and urbanization level in our country can be boosted is related with whether our country can successfully skip past the so-called 'middle-income trap'. So it proposes that systematic implementation of intelligence development and industrial upgrade of a batch of major cities in China incorporates building of intelligent application system in cities, infrastructure and big data platform and formation of a new model for city development characterizing high efficiency, boosted industrial level, guaranteed employment ratio and updated livelihood of city residents. It aims at surpassing 10,000 dollars for per capita GDP before 2020 (when urbanization rate reaches 60%). It intends to grab over 16,000

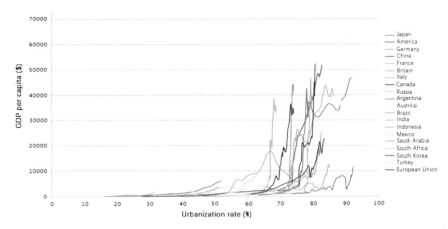

Fig. 1 Development condition of urbanization in relevant countries (*source* Topic Team in Tongji University)

dollars in terms of per capita GDP before 2030 (when urbanization rate reaches 70%), i.e., it intends to opts for development routes in Fig. 2① and timely constructs a highly-efficient and smart economic structure before guaranteeing that per capita GDP steadily rises.

3   Key Construction Content of Domestic Intelligent Cities

Key construction contents of intelligent city cover in-depth and mutually correlative city information network, comprehensive perception on such dominant ecological factors as resources, environment, infrastructure and industry in cities, intelligent decision-making on such rational factors as city economy, technology, culture and management, building of a platform of information-sharing featuring diversified interaction and synergic innovation, boosting of qualities of residents, propelling of employment, spurring of consumption, realization of intelligently configured resources and responses of public services and forging of a people-centered beautiful homeland. Key building contents cover the following suggestions.

3.1   Intelligence of City Construction

1. City economy, technology, culture and management

When it comes to city economy, technology, culture and management, it carries out sustainable ecological economy building, forms collaboration of manufacturing, project and services as well as development driven by synergic innovation, spurs

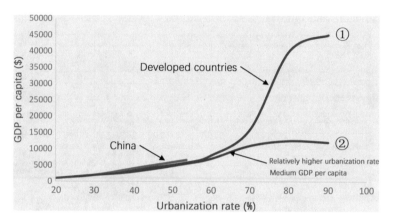

Fig. 2 Demonstration on average development trajectory of ① and ② countries

development of sectors related with local internet of things, seizes upon the new round of economic growth opportunities, clusters multiple sides of resources to carry out core technological research, forms independent innovation capacity, facilitates development of emerging sectors with independent and innovative technologies, promotes formation of innovative culture in cities, constructs public culture service system, enriches cultural life, carries forwards cultural heritage, shapes and publicizes the cultural image of city, propels core value concept to form value cognition, sets up an open, extensive and fair education system, propels exchange and dialogue among different civilizations, realizes innovative transformation and cultivation of innovative talents in fusion of Chinese and western culture, constructs a diversified and open management main body framework involving the whole society, builds a 'nodal network' management form smashing barriers between higher and lower levels, sets up an intelligent management service system with 'integrated high-tech content', puts into force public route, brings into fruition sharing of data, information and knowledge, constructs a public decision—making and management platform involving all people, boosts transparency of social management and realizes social harmony and people's comprehensive development. Key construction contents cover:

(1) Formation of linkage of manufacturing, engineering and services and an innovative network platform and system with synergic efforts by government, companies and scientific research institutions. (2) Formation of a diversified and innovative urban culture system and fostering of an open and fair education system with cultivation of innovativeness as the core.

2. Space organization pattern, intelligent traffic and logistics

It intends to form an intelligent system of 'comprehensive perception—accurate judgment—self-study' in the city and regional planning sector, establish intelligent city space organization model featuring site selection with intelligent layout, city space structure and infrastructure optimization, resource and energy consumption reduction, and with real-time monitoring and response of urban operation. It will establish an integrated mechanism characterizing overall city planning, land use planning as well as traffic planning, establishes an open and extendable platform to shore up intelligent city planning, deeply facilitates research and development of city analysis model and decision-making model, spurs research and development of city planning model supported by big data, establishes a system of city planning and experience decisions as well as system involving public participation and set up an integrated monitoring and analyzing system on basic information in economic and social space in cities and regions.

It intends to establish a comprehensive cloud platform on city traffic information and traffic big data center aggregating space synergic perception on transport state, interactive disposal of information and integrated technologies, realize operational collaboration of different traffic means, significantly boost comprehensive operation efficiency of city transport, realizes intelligent analysis and decision support of city traffic, provides traffic information service on phone mobile terminal, mobile terminal, vehicle navigation mobile TV terminal and o the like which are based on the internet, realizes conversion from public traffic services to individualized traffic services, is competent to provide such services as transport guidance, parking guidance, initiative forwarding of traffic information by capturing traveling habits of passengers, intelligent travel and so forth and establishes a traffic planning and construction system integrating planning decisions and operational response with city intelligent traffic information platform as the basis.

In the sector of intelligent logistics, it will construct a platform with social logistics services that covers infrastructure platform, public information platform and platform of industrial policies to shore up conversion of industrial and commercial companies from conventional operation model to modern operation mode; it will establish a real time and efficient transport system and provide new logistics services via transformation of transport system; it will construct a city platform of logistics services, establishes different levels of distribution network sites and satisfies diversified, mobile and individualized e-commerce development demands. The key construction contents cover:

(1) Construction of a public platform of intelligent city planning and an intelligent monitoring and controlling platform on national urbanization; (2) Construction of a platform featuring traffic perception network and resource sharing, and an intelligent analysis and decision supporting platforms featuring traffic perception network, resource sharing platform and intelligence of Traffic operation; (3) Construction of four major systems of intelligent logistics services respectively a logistics system shoring up commercial services marked by rural-urban integration, an urban logistics distribution system that goes along with e-commerce development, an industrial logistics system that meets demands of transformative development of modern industry and an intensive operation system that caters to international trade.

3. Intelligent construction and home furnishing

In terms of intelligent building and home furnishing, it adheres to "top design, market-oriented, pragmatic, strengthen management" principle, takes the functional requirements as the direction to construct a building intelligent system which can truly achieve security, strengthen management, improve efficiency and energy conservation, establishes a new intelligent building technology system and management system to change the status that discrepancy between the technology and practical engineering application exists in the field of intelligent building, conforms to the needs of building development, so that intelligent building can meet the needs of sustainable development of China's cities and energy-saving emission reduction, establishes intelligent building technology system with China's independent intellectual property rights, breaks the monopoly of foreign technology to make intelligent building as an integral part of "Created in China" strategy. The recommended key construction is:

(1) As regards intelligent construction (public construction): building of an intelligent systematic construction and managing system running through the whole life cycle and construction of intelligent architectural products and data standards; development, promotion and application of advanced intelligent construction system with proprietary intellectual property rights; (2) Smart home (dwelling): building of intelligent home products and data standards; building of guide rules for intelligent home technology and guiding on practical development concerning development and application of home products.

3.2 Intelligence of City Information Infrastructure

1. Information network

In terms of information network, it intends to build city infrastructure of information network characterizing extensive intelligence, openness, sharing, and isomeric fusion, seamless mobility, safety, reliability, greenness, energy conservation, simplicity, transparency, flexible extension and other characteristics, forges the new round of clout compute network platform featuring virtuality of resources, servitization of computers and intelligence of management. It aims at propelling fusion development of 2G, 3G, 4G and higher level with WLAN network, forges a basic information communication network with broadband, flattening and fused nature as core characteristics, propelling co-building and sharing of telecommunication network infrastructure and two-way remaking of broadcast television network, intensifying building of home information network and revving up scientific research and industrial layout of network technologies in the future. It will build a comprehensive information platform shoring up city diversified information, featuring high-performance computation, storage and transport capacity, set up diversified and extensive knowledge and information pilot covering governmental management, public life and industrial advancement and forge a 'guidance headquarter' and 'staff' for operation and management of the whole city. Key building contents cover:

(1) Building of intelligent city information network infrastructure characterizing extensive fusion, high-speed broadband, controllable safety, greenness and energy conservation; (2) Building of an information axis center covering city management, industrial development people's life as well as a high-quality clouding services data center.

2. Geological information infrastructure

With multiple-scaled, multi resolution and multi-source digital city geological space framework as the base, it can boost 'space criterion' to 'time and space criterion' and upgrades '2-D geological information + 3-D visual representation' as '4-D geological information with unified time-space criterion' and upgrades 'postmortem analysis + auxiliary decisions' to 'actual-time analysis + decisions'. It puts into force building of geological infrastructure following time-space standard system, intelligent perceptual linkage and fusion platform, geological space information fusion management platform, building of intelligent analysis decision and service platform that implements infrastructure of geological information. It reinforces actual-time connection and correlation of universal code framework depicting various types of sensors and their interactive connection as well as multisource sensor information under the unified time and space, independent loading and content fusion of multi-source heterogeneous information, high-efficient information renewal in face of changes and other core technologies. Key building contents cover:

(1) Top-level design on geological information infrastructure following framework of intelligent city; building of an unified and authoritative platform on national geological information infrastructure and public services; (2) Building of various information integration systems with geological information as the basis; research on connection and correlation of universal code framework depicting various types of sensors and their interactive connection as well as multisource sensor information under the unified time and space, independent loading and content fusion of multi-source heterogeneous information, high-efficient information renewal in face of changes. It sets up cloud service model of geological information and position.

3. Big data and knowledge disposal

As regards information center building and knowledge disposal in intelligent cities, it should start by building the system structure and framework of intelligent city knowledge center and big system for information disposal, mutually linking to existing information such as digitalized city, informationzied city and smart cities and realizing flexible coupling and linkage of information resources. It should establish parameter system to garner various information (such public services and public safety as water, power, oil, gas, traffic, education, medication, library, network and so on) as well as metadata base with intelligent city as the core factor and its standard system and system of processing technologies; it should research and design supporting technologies for information disposal, knowledge digging release and communication in intelligent city. Secondly, it will plan the layout to make sure that knowledge center layout and city planning system to be integrated; thirdly, it will sort out management model of intelligent city, research and design system and mechanism and policy guarantee system on building and management of information resources in smart city, facilitates fusion of various data and information in intelligent city and dismantles the barrier of 'management wall' for building of a knowledge center and information disposal center. Lastly, with cloud computation technology as the backup and big data management and disposal as the means, it will build a knowledge central pivot covering governmental management, public life and industrial advancement, build a 'guidance headquarter and staff sector' for operation and management of the whole city and construct a new center for scientific and technological culture characterizing participation of residents, government and companies, synergic innovation and lifetime learning. Its key building contents cover:

(1) Key technologies for the overall framework of intelligent city big data center and information integration;
(2) Key technologies for big data management, knowledge discovery and auxiliary decisions.

3.3 Intelligent Urban Industrial Development

1. Intelligent manufacturing and designing

As regards intelligent manufacturing, it will utilize the new generation of information technologies such as cloud computation, ubiquitous network, Web 2.0, embedded system, artificial intelligence, new type sensor and other technologies, carry out study on intelligent manufacturing device, intelligent manufacturing system, intelligent manufacturing service system, intelligent factory, cloud manufacturing system and so on. It aims at realizing automation of physical labor, repetitive work, hazardous and risky health work, bringing into fruition automation of complicated work that surpasses people's controllability, attaining zero-emission green manufacturing and realizing highly-efficient distributed manufacturing so that manufacturing companies return to cities and become neighbors of residents.

As regards intelligent designing, it utilizes the new generation of information technology to carry out product modeling, designing expert system based on knowledge, synergic innovation design system, network component warehouse facing synergic design and manufacturing and so forth. It should not just back up informationization and synergy of product design but also fully mobilize synergic innovation competence of the whole industry and society to spur knowledge accumulation, orderliness, sharing and valid application in an effort to conspicuously boost a city's innovative competence and facilitate economic transformation and upgrading. Its key building contents cover:

(1) Building of a digital intelligent factory;
(2) Building of cloud manufacturing system facing middle and small companies.

2. Intelligent Power Grid and Energy Grid

It will construct a city power network carrying distributed dispatching management equipped with intelligent judging and self-adjustment competence. Integration and use of distributed power generation, notably power generation of renewable clean energy should apply myriads imbedded intelligent devices, distributed computation and communication technologies to conduct actual-time monitoring, analysis and controlling on the operating state of city power grid so that is carries the competence of self-recovery and post-event speedy recovery. Through two-way visibility, customers are advocated, supported and encouraged to get involved in power market and provision of demand response to ensure highly-efficient and reliant power supply with high power quality and reasonable price and to provide backup to application of such technologies of plug-in (hybrid) power automobile, distributed photovoltaic power generation, power storage and energy-saving building. Son the ground of city intelligent power network, it advances city intelligent energy network towards the goal of 'low carbon, high efficiency, hierarchical use, intelligent dispatching and supplementary advantages'. By relying on information communication technology and intelligent data center, it installs intelligent characteristics that help people make decisions and replace power of mankind, shatters the originally comparatively independent energy use system, organically aggregate transformation and distribution of energy resources covering multiple fields covering intelligent fuel, net power, water conservancy, thermal network and materials following the basic rule of energy conservation and emission reduction, establishes a revolutionary model of energy production and consumption, realizes highly-efficient transformation and use of energy source, formulate one-package solution of energy and environmental issues under the unified framework towards intelligent energy network development, constructs an intelligent scheduling platform for energies and forms a brand new backup and guarantee system for city infrastructure by relying on city intelligent power grid construction.

Key building contents cover:

(1) Integration and use of distributed renewable energy for power generation and building of efficient operation and energy efficiency management system in city power network; (2) Building of intelligent scheduling system and service platform with multiple energy resources with conservative consumption and environmental carrying capacity as comprehensive guidance.

3. Intelligent business and finance

As regards intelligent business, it mainly carries out such work as top-level design, safety guarantee, platform building, application, publicity and governmental backup. The Commerce Department, The Ministry of Industry and Information

Technology and other sectors co-formulate implementing planning and specialized planning on nationwide intelligence business, come up with technological standards for intelligent business, work on safety rules for business information, mainly construct public service center on business information, nationwide and globalized operation platform for intelligence business (such as intelligent mall), implement intelligent business model project for large-scaled companies and key sectors, spur e-business to March towards whole-process intelligent business model, deepen and extend applying of intelligence business in industry, agriculture, commerce circulation, transportation, rural and urban consumption and other sectors and carries out financial and tax backup on research and development of intelligent business technologies and platform projects or companies.

As regards intelligent financing, it puts innovation and in-depth application model as the key, marketing as the guidance, social and customers' demands as the center and large-scaled financial outfits as the reliance, tracks the development tendency of internet finance, reinforces in-depth development of financial information, aggregates information of intelligent finance and business, intelligent manufacturing, intelligent medication, intelligent education and other sectors, guides and establishes a unified data platform and finance cloud, realizes data sharing of various financial information systems of different scales and industries, propels representative financial institutions to carry out pilot spots for intelligent financial construction, build protection system for safety level of financial information assets, reinforces safety risk assessment of financial information, works on security monitoring system of financial information, improves capacity to tackle with the guard against network safety events, works on emergency disposal pre-plan for information safety to ensure domestic financial safety and boosts status of intelligent financial industry in modern service industry. Key building contents cover:

(1) Key backup of companies with conditions to build 5 to 8 intelligent e-malls, e-business and financial platforms carrying international cloud; (2) Spurring of commercial intelligence development of large-scaled companies and key sectors covering retail, logistics, banks, security, insurance and so on and building of mobile intelligent business platforms with fame and clout in the industry.

3.4 Intelligent Development of City Management Services

1. Intelligent Healthcare

About Healthcare, it propels application, exchange and integration of electronic medical service records in medical service outfits at all levels and launches unified cloud services for mobile medical hygiene covering acquiring of electronic medical service records of residents, remote medication, unified mobile disbursement for medical services, individualized medical health services and so on. Key construction contents cover:

(1) Actual-time fusion, opening and sharing of medical service records in all outfits;
(2) Building a unified mobile platform for disbursing various medical insurances.

2. Intelligent city environment protection

In terms of city environment, it builds an all-day-long and all-out stereoscopic physical space for city environment perception, constructs monitoring information system for key pollutant, informationized monitor station network for ground environment, high-resolution and hyper spectral environment remote monitoring network and other infrastructure for information perception, facilitates application of such new and high-tech information technologies as internet of things, laser communication, globalized information grid, cloud storage, cloud computation, virtual reality and so forth, launches sharing platforms of environmental information and big data center building, guides and forms nationwide cyber space on environmental information management and optimization of environmental decisions, fully displays city environmental information service functions, builds and aggregates informationized systems for functions covering pre-warning and forecasting of city environment, emergency response, optimization, adjusting, auxiliary decisions and so forth, reinforces sharing and openness of environmental information, improves openness, objectivity and comprehensiveness of environmental information, guides public participation system, improves shoring capacity of environmental message on city environmental management, and facilitates advancement of environmental management model with improved environmental quality, and risk control of environment as the goal before backing up building of an ecological and civilized city. Key building contents cover:

(1) Construction of intelligent perception network, data center and other infrastructure of city environment;
(2) Construction of intelligent decision and comprehensive service platform for city environment.

3. Intelligent city safety management

As regards city safety construction, it sets up city safety index system, incorporates appraisal and planning of city safety into city development planning, carries out building of all-cycle city safety management system covering four aspects namely appraisal, planning, implementing and operation and spurs four conversion concerning building of city safety systems namely conversion from Made-into Made-for, conversion from data provision to information provision, conversion from department to safety information grid and conversion from engineering building to operation services. It configures a batch of innovative platforms for city safety forms independent innovation capacity from research and development to services, reinforces research, development and application of key technologies for city safety such as intelligent video perception, terahertz imaging, flexible sensor, radar acquisition, microwave remote sensing, digital lining technologies and so forth, facilitates and cultivates professional suppliers for city safety operation, probes into operating service model for safe city, sets up a system of city safety information grid, intensifies building of network and information safety, builds secured and reliable information consumption environment, improves guarantee competence on safety of internet information, builds 'national model area for city safety' in regions with comparative advantages, faces the reality that 'prisim door' and incompetence of comprehensive replacing imported network communication devices and technologies, allays possibly-emerging safety issues via mutual constraining of products and realizes independent controllability in the long haul. Key building contents cover:

(1) Construction of a national emergency information grid system for public security;
(2) Construction of safety pre-warning and monitoring system for cyberspace.

3.5 Intelligence of Human Resources in Cities

Intelligence of city human resources is mainly to construct intelligence platforms of human resource during the process of city intelligence on the basis of education, competence, skill, experience, physical power and so on of city employees so that they can be scientifically, reasonably and validly utilized by wall walks of the society, enable employees to realize their maximum values and attain maximum degree concerning contribution of values to the society. For example, such functions as information convergence, outplacement services, guidance services for entrepreneurship, unemployment guarantee services, implementing services for employment policies, occupational guidance and training services and so forth can

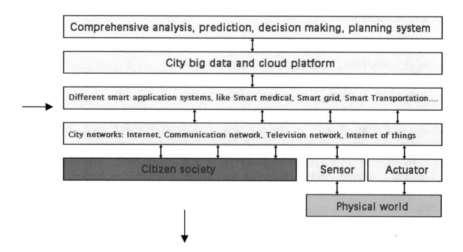

Fig. 3 Key points in intelligent city construction

be aggregated into one through construction of intelligent platform of human resources market and with people as the basis. It helps realize whole-process informationization and networking covering recruiting and job-hunting, information release and so forth. Another example is listed below: building of intelligence platforms of human resources management helps coordinate economics, management science, psychology, operational research, computation mathematics, and relevant professional subjects with intelligence technologies that can realize intelligence of human resources management featuring various characteristics and attain intelligence of decisions concerning human resources management.

In summation, the key points in intelligent city construction, in short, categorized into five levels following routes and structures of its implementation (see Fig. 3). The third level is to conduct in-depth upwards and downwards extension starting from the intelligent application systems in order to realize interconnectivity of 'PHC'.

Chapter 4
Routes and Strategies of Intelligent City Development in China

1 Routes of Intelligent City Development

Intelligent cities in China should put boosting of sustainable economic growth as the goal, ecological civilization construction as the driving force, improvement of people's value concept, life concepts and world concepts as the traction and enhanced safety and sense of happiness of residents as the ultimate goals in order to realize modernization of mankind. Development of intelligent city can be summarized as comprehensive, coordinated and sustainable development prosperity, i.e., sustainable economic prosperity, continuous social harmony and continuous ecological civilization (see Fig. 1).

Formulation of sustainable economic and social development strategies and policies should take into account bearing capacity of environment and social harmony, so the future tendency of cities must be unification of 'economy-society-ecology', i.e., gradual conversion from Fig. 2a–b.

Building of an intelligent city marks a process from separation to integration and from shallowness to profundity. It starts from building of various intelligence systems, gradually carries out top-level designing and comprehensive coordination of all sides, guarantees in-depth fusion of city economy, human resources, city building and planning, city information facilities and city management and services (Fig. 3a, b), and constantly spurs unification of 'economy-society-ecology'. With managers' vision and demands of residents, it dismantles barriers of 'management wall', improves comprehensive level concerning intelligence of city infrastructure, redresses clashes on city development and environment, resource and space, guarantees information sharing and safety, facilitates sound and sustainable industrial development and city building, realizes city management and service systems that satisfy the government, the masse and companies and empowers Chinese cities to have more extensive international competitiveness. With attention going on innovation, it stresses intelligent city operation, ongoing renovation and takes a route with 'building and use' co-existent.

© Springer Nature Singapore Pte Ltd. and Zhejiang University Press, Hangzhou 2018
Y. Pan, *Strategic Research on Construction and Promotion of China's Intelligent Cities*,
Strategic Research on Construction and Promotion of China's Intelligent Cities,
https://doi.org/10.1007/978-981-10-6310-7_4

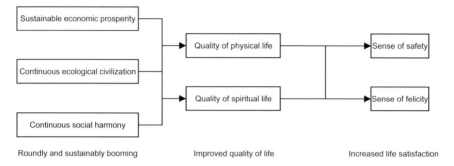

Fig. 1 Roundly and sustainably booming intelligent city models

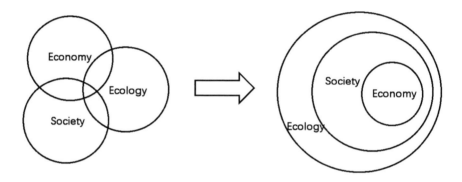

Fig. 2 Roundly and sustainably booming model development

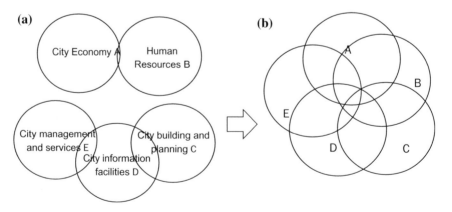

Fig. 3 The sustainable and coordinated development model of each urban system

2 Strategies of Intelligent City Development

The future city construction requires that we carefully imbibe experience and lesson of domestic and overseas city building, carries out city intelligence building, attempt to realize urbanization at 60% and per capita GDP at over 10,000 dollars by 2020 and urbanization at 70% and per capita GDP at 160,000 dollars by 2030 on the strength of its own characteristics in different humanistic environments, geological positions, development phases and social formations. Strategies to build and boost intelligence city buildings cover.

2.1 It Advocates that Cities Spur Intelligence Construction on the Basis of Their Respective Featured Demands

For intelligent city planning, a city should fully consider its demands for local features; shirk such cases as consumption of natural resources during city development, blind city expansion, damage of ecological environment, consistent city pattern vanishing of territory features and cultures. It should be particularly noted that featured heritages left by former generations are passed through generations and form city stories city residents take pride in as well as city brands. We have the responsibility to protect them, pass them down and carry them forward. At the same time such should be guarded against as blind enclosure for city-building, ocean circling for city building and mountain-whittling for city building in the name of transforming an old town and building of new districts without taking objective rules in view.

2.2 It Chooses Pilot Cities to Carry Out Intelligent City Construction

Construction and advancement of intelligent city is a long-term, arduous and grave task. It is a progressive process that calls for advanced pilot spots and constant explorations. For comparatively advanced cities in China, notably advanced cities along the coast. A city should make consideration from its overall development, comply with goals of struggles in the '18th Party's Congress' and requirements in the national strategies, come up with special development planning for 'intelligent cities in China 2025' from the national level, adopts advancement of pilot cities, drives planes with spots until gradually radiating over all domestic cities, and incorporates building and planning of already authorized 'smart cities' into the planning.

2.3 It Takes Intelligent City Construction as the Basic Platform to Spur in-Depth Fusion of 'Four Modernization'

With construction and advancement of intelligent city as a major seizure for economic advancement in our country and a basic platform for in-depth fusion and development of industrialization, informationization, urbanization and agricultural modernization, reinforces transformation and upgrading of traditional agriculture, industry and service industry, cultivation of emerging industries, innovation of branded products, protection of intellectual property and other measures in cities, facilitates coordinated development of all sectors in the region or throughout the country, averts severe industrial convergence and repetitive investment, realizes boosted employment, shortened gasp between the rich and the poor, drives forth consumption, improves livelihood and ecology, guarantees social just and other national goals and trails a new route featuring Chinese characteristics that takes a lead in the world.

2.4 Safe Convergence and Sharing of City Big Data Is Guaranteed with Government as the Dominance and Institutions and Companies Getting Involved

Reasonable use of big data will trigger huge fortunes that would not just boost quality of city management services but also facilitate people's clothing, food, accommodation, travelling, work and study. It is also new and major production factor that has become key building content all over the world, in particular advanced countries with America as the representative. Launching of data center and knowledge center building should become a major priority for building and advancement of intelligent city in our country. On the one hand, it calls for governmental dominance and company participation and the on the other hand, it requires that we dismantle information barrier, information isolated land, shirk repetitive building and guarantee safe convergence and sharing of big data in cities (see Fig. 4).

2.5 Facilitation of Standard Development of Cloud Services (Procurement of Cloud Services) and Other Information Service Industry

For internet economy, procuring of cloud services (Mao 2014) is tantamount to a big reform during 'all-round contract' period (household contract system' in rural

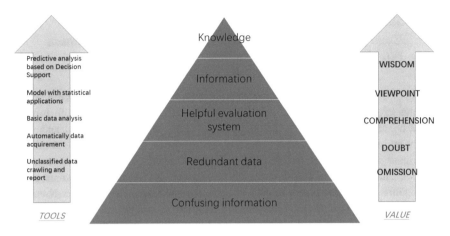

Fig. 4 Big data accumulation and use *Source* PPT of Academician Xu Kuangdi

areas back then. Procurement of cloud services refers to buying of cloud storage and cloud computation services of data. It refers to relations of commercial activities incurred with residents, companies and governments, buying objects, and cloud service companies that provide cloud storage and cloud computation services as the subjects. So in both ideology and actions, we should dispel people's worries on procurement of cloud services, legitimately foster, forge and regulate procurement of cloud service market, boost network rights and safety guarantee capacity, give display to its 'guidance role and 'realization' role of business model and facilitates speedy, sustainable and sound development of online economy.

2.6 Intensification on Technological Standardization Related with Intelligent City

Standardization is the major foundation for building and advancement of intelligent city. We should establish a standardized facilitation and coordination mechanism with relevant commissions and ministries, industries and associations involved, launch unprecedented improvement on operation model of existing city outfits, constantly reinforces self-learning and perfection of standards, eventually realizes mutual fusion, mutual exchange of data, and proneness of integration of functions to satisfy interests of all sides. We should also reinforce dominance and internationalization of standardization building and guard against abuse of standards or being instrument to seize monopoly by interests' bloc.

2.7   Set-up of Intelligent City Appraisal System that Can Embrace All Characteristics and Different Development Phases

Set-up of an intelligent city appraisal system is not aimed at comparing with others. It helps city government and residents to comprehend the huge profits brought about by city intelligence so that they would cater to local conditions and more viably and practically construct cities they dwell in. So it is needed that a tolerant intelligent city appraisal targeting at different development stages and carrying various features be built and such should be aligned with as that development stages and features of different cities should be taken into account under the comparatively unified national system by highlighting national dominant role and displaying subjective initiative of local places, attaching importance to alignment of Intelligent city construction and local development demands and stressing actual propellant role of performance of Intelligent city construction, stressing participation and appraisal of professionals and giving play to participation of social and marketing power.

Reference

Mao, G.L. 2014. Purchasing cloud services: A big systematic revolution. *Informatization Construction* 3: 24–26.

Chapter 5
Suggestions on the Intelligent City Construction and Promotion in China

On September, 30, 2013, President Xi Jinping pointed out that, 'a new round of scientific and technological revolution and industrial reform will form historic intersection with the development means of quickened transformation of economy in the near future that will provide a precious major opportunity to realize innovation-spurred development strategies.' City intelligence building is a major opportunity to innovation-spurred advancement that can fend off issues incurred by 'smart city' building recommended by international IT titan. Therefore, it is suggested that we initiatively seize upon big trend and opportunities of global city intelligence, take city intelligence building as a basic platform for new urbanization, upgraded industrialization and in-depth informationization, as major basis to realize new regional development, as major seizure for economic and social development and progress of city civilization and potent force to realize 'Chinese dream'. In view that domestic cities have higher governing ability, city intelligence level in China is expected to reach a new world level. Concrete suggestions are listed below.

1 Formulation of Special Development Planning for 'City Intelligence in China, 2025'

Construction and advancement of intelligent city involves prosperity of a country, wealth of people and national core interests of safety. It is suggested that the Central Government reinforce leadership, reinforce special development planning of 'intelligence of Chinese city, 2025', call for concerted efforts of politics, industry, academics and research and incorporate authorized 'smart city' planning in. The planning covers formulation of '13th Five-Year-Plan' for Chinese city intelligence as well as special development planning of 2020–2025. The planning contents cover: classical city planning, selection of classical cities in coastal area, central and western areas, North China, Northeast China and so on with pilots going ahead and

© Springer Nature Singapore Pte Ltd. and Zhejiang University Press, Hangzhou 2018   61
Y. Pan, *Strategic Research on Construction and Promotion of China's Intelligent Cities*,
Strategic Research on Construction and Promotion of China's Intelligent Cities,
https://doi.org/10.1007/978-981-10-6310-7_5

plane driven by spots; major special project planning and special planning of technologies such as comprehensive analysis, forecast, decision and planning, big data technology and cloud platform, city information sharing and safety, various intelligence applications (intelligent power grid, traffic, medication, hygiene and so on), new city network (internet, communication network, remote network, TV net, internet of things and so on), sensor and executors, infrastructure, social guarantee system and so on; IT hardware and software; industrial planning such as industrial upgrading and remolding, development of emerging industries, brand manufacturing and so forth.

2   Reinforcement of Talent Cultivation

By combining with intelligent city Construction, it systematically boosts knowledge, competence and qualities of residents, cadres and entrepreneurs. It correspondingly optimizes layout of higher education and scientific development orientation of the country, combines with national talents projects to quicken discovery and cultivation of a batch of city planners, management experts, high-level scientists, experts on data science and safety, engineering technology experts and so on that cater to new technology revolution tendency. At the same time, it stresses and clusters power to cultivate a large batch of maintenance engineers and technicians who know both theories and practice. By organizing and implementing national key projects, it clusters and forges a batch of high-level research and development talents. By depending on demographic dividend, it gradually turns to knowledge and talents bonus. Moreover, it also vigorously carries out interdisciplinary research featuring 'knowing, developing and utilizing human brain' so that intelligent city construction and advancement sustainably and soundly evolve in depth and width.

3   Listing of Construction and Advancement of City Intelligence as a Top Leadership Project in Cities

City intelligence construction is a long, complicated and systematic project that requires sustainable and sound facilitation. We should guarantee that first-level leaders in the city take account and all functional sectors and all walks of the society take part, carry out strategic planning and top-level designing, follow overall planning and yearly planning on city intelligence to organize implementation and do not shift with replacement of people. At the same time, it would display the dominant role of the Central Government and provincial government, intensify work guidance and scientific appraisal, guarantee sharing and safety of city information, sort out policy resources, back up city intelligence construction, correct work errors, complete and work on mechanism and bring out long-term maximum interest of the country, local places, residents and companies that can go through text of the history.

Appendix
Report Abstract of Varying Tasks

Abstract of 'Economy, Science, Technology, Culture, Education, and Management of Intelligent City
A Development Strategic Research in China'

General View on Intelligent City Development Strategies

Informationization, globalization and urbanization are re-shaping modern society and city development, but speedy urbanization process is bringing about a string of issues for city operation and development such as ecological deterioration, shortage of food, scarcity of energy, financial tsunami, and terrorism and so on. Those issues are constantly sprawling mainly because cities are yet to develop into a self-adjusted system with sustainable development. Therefore, city development must take a road of intelligence, tolerance and sustainable development in the future.

Findings through early-stage research indicate that for the moment, the country as a whole is actively building intelligent cities, but the bulk of cites only have a vague idea on prospects and strategic goal of intelligent cities and leading cities in building of intelligent city are short of overall plan and top-level designing that stress technologies and overlook management in concrete practice. It would necessarily lead to a string of issues covering information isolated idea, project scattering and so forth and deviate from the initial intend of Intelligent city construction featuring 'comprehensive perception and connectivity'. It makes intelligent city construction linger over the stage of informationized building and falls short of forming concerted effort for intelligent technology and application.

In summation, we reckon that the prospect of intelligent city should be a comprehensive, coordinated and sustainable development model on economy, society and ecology before boosting quality of physical and spiritual life of residents and give a sense of safety and felicity to residents.

© Springer Nature Singapore Pte Ltd. and Zhejiang University Press, Hangzhou 2018
Y. Pan, *Strategic Research on Construction and Promotion of China's Intelligent Cities*,
Strategic Research on Construction and Promotion of China's Intelligent Cities,
https://doi.org/10.1007/978-981-10-6310-7

   To be comprehensive and sustainable regarding the aforementioned three aspects, strategic framework of intelligent city is needed to orderly realize relevant prospects and be progressive so as to bring about the most valid use with limited finance and resources.

   The framework of intelligent city is an entire process that calls for top-level designing to ensure entirety of the framework, for example, it starts from transfer of economic growth means, evolves from resource-driven model to innovative-driven reform, and transfers from low-value, high-energy-consumption and high-pollutant and extensiveness to sectors featuring high added value, low energy consumption and pollution and intensiveness in a bid to realize transformation and upgrading of industries. Moreover, scientific and technological backup quickens transformation and upgrading of industries and spur central round concerning transformation of economic development means. We should reinforce the role of science and technology on backing up economic transformation, vigorously foster innovative subjects, and ask for more talent fostering and education backup in an effort to construct a regional highland for innovative talents. We attempt to dig into the framework of intelligent city strategies from the perspective of functions and categorize it into five sub-strategies namely technology, economy, talent and education, culture and social management.

Strategies on Economic Development in Intelligent Cities

Against the objective setting that urbanization process gathers speed, we probe into economic and management development demands in domestic intelligent cities by starting from three wagons that drive forth economic development in cities and analyzing effect of investment, consumption and export on economic development in intelligent cities. We further indicate that the key to facilitate economic advancement in intelligent cities lies on transformation of development means, facilitation of upgrading of traditional industries, fostering of emerging industries, quickening of industrial fusion and that the key to boost management capacity in intelligent cities lies on transfer of management concept and innovation of management means.

   Intelligent city construction in our country calls for reinforcement in building competence in the following manners namely competence of coordinated planning and development, boosting of development competence of city intelligence industries, improvement of innovative development competence of traditional sectors, elevation of public service capacity in cities, comprehensive perception, intelligent disposal and construction, intelligent peer, endogenous development, high integration, full aggregation and independent innovation competence. We further opt for such classical sectors as intelligent logistics, intelligent transport, intelligent device, intelligent medication and so on for in-depth analysis and discussing the role displayed by classical sectors on boosting economic development and management level in intelligent cities.

The core of economic development in intelligent city is to put physical space determined by intelligent city planning as the basis and to entrench people-centered new city roads. By depending on building of cyberspace, it vigorously develops modern information technology industry and information service industry and facilitates intelligent upgrading of traditional industries. By revolving upon space building, it vigorously develops people-centered modern service industry. By relying on development of intelligent medication, transport, education and pension, it boosts mental level of city clusters before driving forth formation of innovation-driven new development model. Against the marketing ambiance of complicated and variegated economic, political and social development pattern, 'made in China' will march towards globalization, informationization, virtuality, intelligence and greenness while 'engineer in China' and modern service industry will deploy high-end intelligent software, internet of things, internet, cloud computation, big data and the new generation of information technology to march towards greenness and intelligence. Such has been determined that driven by the dual wagon of 'China projects' + 'made in China' and with modern service industry as innovation propellant, it is to realize status of carriers and behavioral selection of interest bodies, probe into modern service industry driven by 'China projects' and 'made in China', in particular the route selection characterizing fusion of information industries and analyze concrete mechanism and linkage effects with interaction of varying factors.

Strategies on Scientific and Technological Development in Intelligent Cities

The topic team comprehends building of intelligent city from the perspective of 'physical space' (first space)-cyberspace (second space) and social and awareness space (third space'. Intelligent city construction with development and application of the new generation of information technologies is mainly clustered in forging of the second space and building of the first space while basic focal point of scientific, educational and cultural building lies on expansion of the third space and building of more in-depth and long building topics for Intelligent city construction. Therefore, scientific and technological development strategies of intelligent city should focus on three aspects namely (1) facilitation of building of economy, society and ecology of intelligent city via valid utilization of science technology; (2) perceiving scientific and technological innovative competence as long-term strategic goal in the intelligent city; and (3) mutual coordination and joint development of technology along with culture and education. In this way, scientific and technological development in an intelligent city should be built on reinforced valid management and use of knowledge resources in cities with cultivation of city innovation as the core. It should aggregate government, colleges and scientific outfits, company and social public, comprehensively foster innovation on science, technology, economy, culture and society and converts economic and social development in cities to reliance on the path characterizing scientific and technological development and innovation, energy conservation, environmental protection and sustainable development with mutual facilitation of three aspects namely scientific and technological development, industrial cultivation and Intelligent city construction. As regards specific development orientation of science and technology, it revolves upon the overall development goal of intelligent city, should stress scientific and technological development of information technologies, combine with concrete features in specific cites to pertinently advance key science and technology and cultivate corresponding scientific and technological industry. In terms of scientific and technological development, particularly when it comes to application of scientific and technological outcomes, it should attach importance to the 'double-edge sword' effect of science and technology and fully consider the adverse impacts or hidden dangers of science and technology on society, economy, ecology and environment.

Strategies on Cultural Development in Intelligent Cities

Cultural strategies in China should fully aggregate various cultural resources to form comprehensive social and cultural development. Therefore, we should start by proposing three major strategies for cultural development namely (1) Forging of city cultural image and of soft power based on ecological civilization; (2) Aggregation of all-out development of value concept and cultural qualities of traditional culture and western culture; (3) Formation of innovative cultural atmosphere with interaction between man and the society as well as development of cultural innovative output fusing culture and technology. Then, it expounds ecological civilization and reflects essence and cultural connotation of intelligent cities. Moreover, in view of fusion of economy, technology and economy, this chapter analyzes ways and routes that science and technology spurs advancement of cultural industry and technologies guided by culture and further states that intelligent cities make innovation and manufacturing as a behavioral means of the society and cluster and make innovation and manufacturing part of city work and life via speedy promotion and sharing of new knowledge and ideology and innovation generation system of internet platforms.

Strategies on Talent and Education Development in Intelligent Cities

Knowledge is the first productivity while education is the basis for knowledge creation, communication and application. By boosting quality and knowledge skills of educatees, residents get improved in individual values and have their social values enhanced. Sustainable harmony and transformation of economic growth means of the society has to depend on development of education.

Intelligent city education backed up by the internet has been put into effect in many countries. The practice is mainly reflected in building of infrastructure of public network education, building of platform of network education resources and provision of handy online education. Thanks to public online education facilities and diversified education resources, students can receive education anywhere and at any time, which largely boosts convenience of study and people's opportunities of receiving education. In the meantime, high quality and low cost of network education avails the whole society.

This project team reckons that the strategic route of intelligent education is to play out the routes of intelligent education, construct 'education clout' that shores up intelligent city development, set up an extensive and intelligent education platform and validly display regulated tailored advantages of online education through professional education services incurred by global integration on system, resources and culture; the strategic vision of intelligent education is to make sure that each and every resident acquire individualized study routes, get adjusted to future professional knowledge and skills of economy via study of society and boost education quality at all levels at low cost and with high quality. It aims at better education skill, more flexible labor force, innovation-spurred intelligent synergy, endogenous intelligent city and realization of 'three comprehensiveness, two combinations and one improvement', i.e., 'building of an education system for all-people education, all-space education and all-factor education', coordination of education development and core value system of socialism combined and that with economic transformation and upgrading in combination; building of a sounder and more sustainable 'social mental space' via mental education, boosting of skill education and elevation of people's quality, social responsibility and patriotism before eventually realizing physical pattern of the whole society and all-out reshaping of spiritual ethos, improve sustainable economic and social development and step out of the mid-income trap with overall strength.

Strategies on Management Development in Intelligent Cities

In-depth analysis on future city development tendency and demands of city management denotes that management strategies of intelligent city should convert from 'passive adjustment' to 'initiative manufacturing'. Viewed from demands for city management in Intelligent city construction, it mainly hails from in-depth changes from three aspects that propose new demands to city management: first, speedy expansion of city population quickens pressures facing city management; second, residents have their requirements on city life quality improved and hence raise their requirements for quality of city management; and third, with advancement of internet of things, cloud computation and other new technologies, new topics are provided to means of innovative city management. These three aspects mark boost of city management from physical space to cyber space to space management demands of social mental space. The topic team reckons that the key of intelligent city management is to coordinate scientific and technological development of physical space and cyberspace, put upgrading of social mental space building levels the objectives and takes the vision of intelligent city management as to construct civilized, democratic, ecological and efficient city ecological system and intelligent operation and management system. The purpose of intelligent city management is to construct four major systematic platforms namely information perception system. Basic data system, decision supporting system and cloud service systems and boosts people's social welfare, resource bearing competence and operation service level via innovation of three systems namely administrative management system, public service system and infrastructure operating systems.

This topic team proposes use of coordinate effect of dynamic change and optimized designing to facilitate congruent development of intelligent cities. It deploys the evolution system marked by 'dynamic change' to manage 'people' in intelligent city and utilizes optimized designing and controlling system to manage 'objects' in intelligent cities. Combination of the aforementioned two systems to Intelligent city construction would not just reflect characteristics of autonomous evolution and artificial designing in management activities, but also provide a valid solution to a complicated issue, i.e., it can redress issues with scientific designing and optimization that can be pre-arranged and resolved with scientific methods. In reverse, it would give play to people's creativeness.

Development Planning in Intelligent Cities

When looking at the evolution process of famous cities in the world as a whole, formation of featured advantages in a city takes a long time. Intelligent city construction is a long-term and complicated process involving multifold aspects, so it asks for long-term overall planning. Therefore, intelligent city construction should focus on the long-term trait when macro strategic planning and layout is the key. The prospect of intelligent city is sustainable economic, social and ecological development. Its realization process should be progressive before validly utilizing limited finance and resources.

Internal interaction exists among all strategic planning, so synergy among strategic planning should be considered. For example, intelligent city construction primarily relies on engineering technology to realize city connectivity but city building is for the benefit of facilitating residents. Therefore, intelligent city construction can't overlook humanistic care in cities that asks for supplementation of cultural and educational planning. We will take Singapore, Amsterdam, Paris, Beijing and Dalian as an example to introduce measures and experience on how they make intelligent cities.

Abstract of 'Spatial Organization Pattern and Traffic of Intelligent City A Development Strategic Research in China'

Information technology throughout the world is entering a new fast development stage. Against this setting, information technology and city development shows tendency of 'fusion'. The so-called 'smart city' has been what a growing number of cities both home and abroad are after. The CAS has proposed that we should quicken promotion of 'intelligent city' building and make city show sustainable development in an intensive, green, hospitable, secured and sustainable way via in-depth fusion of urbanization, informationization and industrialization. It is both related and significantly contrasting to the concept of 'smart city' proposed in the west.

Intelligent technology alone can't bring about an intelligent city. Industrialization is a major milestone in the development history of human civilization, but accompanying technological progress brings about ample fruits and unprecedented insecurity as well with environmental pollution, social conflict, Warcraft and technologies going hand in hand. Informationization has entrenched a new milestone for human civilization advancement, yet if it is short of an explicit strategic guidance, technological development might incur more issues. Therefore, during the building process of intelligent city, we should come up with clear strategic vision and route designing so that it can serve people's boosted life quality rather than subjecting people and city to technologies.

As one of the results of 'study on intelligent city construction and promoting strategies in China', a major consultancy program in CAS, this book mainly centers upon building of entity city and flowing outfit of entities. Then how should we cope with ways that macroscopic tendency of information technology development on organizations with varying functions and how to deploy information technologies to improve city building and operation performance in a bid to construct a fundamental space for PHC fusion development? This book will demonstrate its statement from three professional sectors: the first sector is on spatial organization of intelligent city. It discusses impacts of technological progress on organization of activities of human society before digging into how to alter a city's functional organization and how to consult to information technology to redress city issues that can't be traditionally worked out. The second sector is on traffic organization of intelligent city. It digs into how technological progress reshape traffic system in cites and it helps individual travelers to more wisely plump for traveling means and routes and better arrange layout of transport facilities and operation of organization traffic system. The third sector is on logistics organization of intelligent city. It discusses how fusion of virtual flow and entity flow thoroughly alter the process of physical circulation before truly boosting logistics efficiency, bringing down logistics cost and serving updated environmental sustainable trait and people's living quality.

Intelligent city construction is an emerging sector. The world as a whole is exploring it with variation in both objective and behaviors. The international experience reflects characteristics in vision building, technology selection and building model, but it shares consensus in alignment with technological revolution tendency, servicing of sustainable development and improvement of city operating performance. In carrying out intelligent city construction, we should acquire broad international horizon in terms of space dimension but have in-depth revolutionary vision in time dimension. We should form clear knowledge on the historic development stage and have enough understanding on local development background we are in.

Since the 18th Party's Congress, our country has proposed the major decision on taking a new featured urbanization road characterizing people-centeredness, synchronous four modernization, optimized layout, ecological civilization and cultural heritage that thoroughly reshapes the old urbanization route and city development means. New urbanization carries out coordinated planning on the society, economy and environment, but it should be put into force in entity space, city residents, objects and other entities. Against this background, intelligent city construction should boost city development performance via improved information technologies, replace fluxion of people and objects in the physical space with a part of information flowing in the virtual space and establish a spatial organization space, traffic operating pattern and logistics operation model matching the informationization process. Moreover, intelligent city construction should improve decision-making process of these entity activities via information technologies so that it can more scientifically, rationally, efficiently and better serve human needs.

In recent years, intelligent city construction in China has been roundly implemented. The country has launched myriads supporting and guiding polices and on the local level, it has been doing plenty of positive explorations. However, in contrast to the objective of intelligent city construction, many issues still remain:

First, it lacks theoretical guidance and overlooks the essence of informationization. Featuring high instrumental and functional nature, it attempts to interpret intelligent city as an instrument to improve governmental policies and management efficiency. It lacks knowledge on subversive impacts displayed by the new round of informationization on physical space planning and building. On the ground that it can't comprehend future city organizational for, it stops short of intelligently guiding and corresponding to city development process.

Second, it is short of systematic designing and overlooks roles of market and residents and the urge of the whole society on improvement of production and livelihood via information. An intelligent city by no means amounts to affairs undertaken by the government alone, it should be the city future jointly made by the society as a whole. In the traffic and logistics organization sector, some cases of coping with the information era by traditional management ideology have shown up that would, to a certain degree, snuff out development of innovative activities.

Third, it lacks clear goals and becomes the appendix for technological experimentation and business operation. Though many cities have proposed the objective of intelligent development, yet they are short of sustainable contents and would be guided by technologies. It would constitute a major trap for Intelligent city construction and makes local government be positioned as the major buyer of intelligent city system and would, on the other hand, lead to deviation of Intelligent city construction from the goal of intelligent development and hence stops short of servicing demands of transformational development in cities.

Fourth, it lacks information supply so that intelligent city construction is groundless. Though theoretically put, new informationization would bring about mass information, yet institutional partition and the awareness of closure would pose as a great challenge for boosting the initiatives. In relevant decision-making sector, scientific rationality still has a long way to go. In the sector of marketing and residents, they still face challenges.

Intelligent city construction should maintain high perspectiveness; fully comprehend remolding of informationization on production and livelihood. We should foresee how this remolding process transfer up and down of varying function sand changes on the principle of site selection and propose coping strategies by stages in city planning, building and operation so that Intelligent city construction is an opportunity rather than a challenge. Research reckons that the most direct result of information technology development is the gradual fusion of entity space and virtual space. For a start, many human activities with physical space as the carrier have gone through transfer to the virtual space. Secondly, human synergy based on space is replaced by synergy based on internet. It has thoroughly altered the clustering condition of city residents and companies and incurred high freedom for site selection for both production and livelihood. Thirdly, virtual space activities would bring about changes of network structure in the spatial activities. Research has anticipated that intelligent technology development will incur the following impacts: first, mixture and segmentation of living community; second, scattering and synergy of industrial production; third, individualization and centrifugation of infrastructure; fourth, mobility and sharing of public services; and fifth, potent polarization and centrifugation of regional development. Under this setting, it can be regarded that the new round of informationzation would bring about revolutionary changes on urban and regional development and organizational means of traditional city space, traffic and logistics. If old corresponding means are still adopted, organization of physical space and entity fluidity will fall into endless disorder.

Therefore, we should distinctly clarify the basic principle of intelligent development in cities when boosting organization of space, traffic and logistics organization. In this research, we have proposed that intelligent development of cities should be carried out on the ground of appraisal of impacts exerted by information technology development. By fusing technology to city development process, it realizes minimum consumption of resources and maximum social output. The core of intelligent city construction is five principles:

First, respect of value concept of environment. Extensive use of information technologies alters the dauntless decision-making means and aligns city development with environmental development.

Second, understanding of systematicalness and clustering of city development, building of multilayered compound organizational means of city functions characterizing spatial recombination, stereoscopic recombination and interleaved recombination and changing of deeming city building as an old route of mutually separated spot-like growing process;

Third, building of a whole life-cycle measurement and putting the answer to satisfy city development and redress city issues into the whole process of planning, designing, building and operating rather than starting from a specific round and taking reciprocal impact of different rounds into consideration.

Fourth, conversion from 'carrier of flow with form' to 'fixation of form with flow' in terms of decision-making logic. Decision-making means is no longer based on adjustment on varying flowing processes in a city after static blueprint is completed, rather, it predicts movement scenes of various flows based on information technology methods prior to planning and constructing before launching corresponding physical space and physical facility layout;

Fifth, minimum energy consumption. Intelligent planning and building helps bring down various consumption of resources during the process of city building and city operation.

Space organization of intelligent city should correspond to huge challenges facing city development under the condition of speedy development of urbanization such as dicey spatial demands incurred by dynamic demographic structure, matching city function brought about by unreasonable city growth, environmental pressure-bearing incurred by population growth and surge of demands, fragility of economic backup brought about by changes of domestic and international situation, instability of social status incurred by unbalanced distribution of space resources and improvement of rights awareness. In the city development and building process, main causes giving rise to these issues cover insufficient, opaque and disconnected decision-making subjects and marketing information and insufficiency concerning decisions on boosted application information and selection competence. By referring to improvement of information supply and application competence, breaking through traditional restriction of space distance, synergic development of virtual space and physical space and shattering low adaptability of city building with high adaptability of city operation, efficiency of city development might be improved, space layout during the urbanization process might be optimized and city operation quality will be improved.

To realize this goal, we should start by ways on how intelligent technologies spur transformation of city development means. It requires that we break through the barrier of physical ideology during the city planning and building process, replace tangible facilities with intangible services and low-efficient physical supply with efficient operation organization and transfer concentrated core form with scattered network forms. It requires that we proactively consider new routes to realize varying demands of residents when carrying out planning layout of intelligent city and take varying city services based on information technology as part of varying functional organizations in cities and truly carry out layout with updated city operation performance, handy life of people and ecological and environmental friendliness. For example, high-quality online medication can alter the status of huge gap on medical quality, overloaded people in some hospitals and paucity of people in grassroots hospitals. Correspondingly, the fundamental principle on layout of medical facilities will witness major changes. Aggregation and release of traffic information will more validly configure operation of traffic facilities, spur residents to make wiser choices on grounds that residents are aware of it, enable remote regions to incorporate their unique edges to modern economic system with ubiquitous network and improve the lopsided status that economic development is mainly concentrated in a handful of regions.

To realize this goal lies on how to utilize intelligent technology to realize transformation of city planning means. It requires that cities actively apply new information technologies to make decisions be scientific and democratic during the planning, management and operation process and realize modernization of governance system during the process. In the traditional development model, decisions on city planning are usually based on subjective knowledge such as ideal and experience of decision-makers and advisers whereas standards used by decision-making is usually based on the social average value and overlooks disparity of people. This situation is, to a large degree, attributed to shortage of sufficient data, reliable model and transcendent judgment. By referring to a wide base of cases, this book reckons that with remote data, massive data, open data, mobile data and new data sources popping up as well as constant advancement of big data, cloud computation and other technologies, city planning can consult to these means to gradually get to know the internal rules of city operation and facilitate democratic decision-making process via comprehensive perception, accurate judgment, proper response and sustainable study.

The goal of traffic organization in intelligent cities is to realize convenience, efficient operation, safety, reliability, energy conservation and environmental protection of traffic system. Such issues as traffic congestion, difficult parking and traffic pollution accompanies city development process and its core lies on deviation between land use and traffic system organization and secondly, it is attributed to feeble management and low selection due to shortage of traffic information. Collection, analysis, sharing and supply of large-scaled and multisource traffic information would, to a large degree, remove difficulties facing traditional traffic organization model.

The breakthrough points of intelligent traffic organization cover: first, realization of minimum traffic demands via integration planning of traffic and land use and improve traffic efficiency, bring down traffic energy consumption and pollution from the very basis. We should comprehensively consider synergic mechanism between development mode and traffic supply facilities, set up self-study, self-organization and response system of traffic planning, conduct response and revision of the planning initiatives based on building and operation data and attain rationality of decisions on traffic planning and building. Second, we need to set up synergic decision-making, synergic controlling and synergic operation of multiple traffic means. We should realize round coordination among different transport means and between traffic and other city systems in such rounds as perception of traffic state, information analysis, disposal and integration, information release and application so that we can fuse complicated city systems otherwise separated. Third, we should bring to fruition most optimal management and controlling on grounds of information support, set up flexible configuration between demand and supply, redress pressure of road, parking and operation and timely cope with emergencies. Fourth, we should realize the most optimal decisions of travelling individuals via information release. By referring to actual-time provision system and forecast and analysis system of public traffic, road status, management information and facility supply, residents can opt for their travelling time, traffic means and traffic routes and overcome bottlenecks of the system with people's initiatives on the basis of the actual condition of the operation of traffic system. Fifth, we should propel individualization of traffic outfits via comprehensive mastery and precise analysis of individual demands for traffic.

The key of realizing city traffic intelligence lies on acquiring, analysis and openness of information. This book reckons that we should establish traffic perception facility aggregating sky, air and land, realize comprehensive mastery on traffic information, fully acknowledge status and rules of traffic flow and operation and comprehend comprehensive integrated application.

Changes of economic and social outfits and people's living styles have proposed more new requirements and new development possibilities for logistics organizations. Internet of things, public logistics information platform, macro and micro logistics network planning technology, optimized technology on operation and management in logistics companies and others have gradually entered logistics sector and new business models are reshaping traditional logistics sector. The objective of logistics organization in intelligent city is to carry out reorganization of occurrence process of logistics activities by consulting to the route of informationization in a bid to boost use ratio of resources in city logistics activities, improve productivity level in companies, bring down carbon emission amount and enable it to serve sustainable development in the city.

This book proposes that intelligent logistics organization should center upon redressing of three main issues: first, it is to probe into solution of diversified logistics demands by consulting to business mode innovation based on informationization; second, it is to realize precise management of transport companies, optimize and reduce consumption and logistics center via reasonable adjustment of vehicle and logistics distribution routes and expand network and boost efficiency via intercompany synergy; and third, it is to coordinate clashes between logistics activities and city development, reduce its impacts on city development by improving structure of freight and distribution system and encourage joint distribution in the region to improve traffic congestion and reduce resource consumption.

Logistics system is the bloodline of sound and robust economic activities and its operation features cross-regional, cross-company and cross-industry traits. In practice, intelligent advancement of logistics system calls for concerted efforts of both the government and the market. First, we need to optimize environment for logistics policies and unshackle institutional barrier and block barriers constraining its development; second, we should reinforce infrastructure, propel integration of infrastructure and logistics infrastructure, improve aggregation of existing resources and synthetic use of facilities and intensify planning coordination and functional aggregation of newly-built facilities; third, we should reinforce interindustry communication, spur interaction between logistics companies and producing and business companies facilitate information interconnectivity between logistics supply and demand market and the government and improve communication and coordination system between companies, sectors and departments; and fourth, we should spur research, development and socialized application of logistics technologies.

Lastly, this book reckons that intelligent development is a crucial means to bolster featured new urbanization path in China that needs to be jointly propelled at the national and city level. At the national level, top-level design is needed for development of intelligent city: first, we should pay attention to and viably intensify relevant standard system and standardization building; second, we should come up with strategies to boost major infrastructure at the national level; third, we should enact data synergy sharing mechanism; fourth, we should establish and work on information network safety mechanism; and fifth, we should put into force demonstration project on planning pilots. At the city level, the following initiatives should be actively launched: first, it should propose intelligent strategic instruments and its selection means during the city building and development process; second, it should propose building strategies and measures for city information infrastructure; third, it should propose development strategies for such sectors as city safety, intelligent traffic, public services, electronic governmental affairs and so on; and fourth, it should propose building strategies for such intelligent platforms as information sharing, decision-making backup and industrial services.

Extract of 'Smart Grid and Intelligent Energy Network A Development Strategic Research in China'

In the 21st century, high-carbon development model after industrialized revolution is up against great challenges along with draining chemical energy and increasingly conspicuous global climate issues. Low-carbon economic development model has shown fundamental reform on existing production and living means that has incurred profound impacts on the global politics, economic, investment, trade, consumption and energy development pattern. In this revolution, energy industry is the most dominant round in the whole variation. According to estimation of American Academy of Engineering, electrification marks the grandest achievements obtained by mankind in the 20th century, but power system that is closely related with us is facing more and more challenges. In face of requirements for worldwide energy and environmental protection and sustainable development, we need to vigorously develop low-carbon technology, facilitate efficient energy conservation technology, actively develop new energy and renewable energy and reinforce intelligent power net and intelligent energy net building. All have become major development strategies to address energy issues in the 21st century.

City Intelligent Power Grid

When looking at electric power supply network as a whole, the power company has been putting emphasis of intelligence on power generation and power transmission system all along so that it is higher than power distribution and use in degree of intelligence. But in contrast to the power generation and transmission round, cooperation by power transmission, power use, electric power companies and terminal users is comparatively feeble, which seriously affects overall performance and efficiency of the system. In the meantime, some issues urgently in need of being redressed show up in city power grid development both home and abroad, for example, reliability of power supply, power energy quality as well as connection and demand-sided response of distributed energy are concentrated on the power transmission net, so development of city intelligent power grid is a topic urgently in need of being redressed that has drawn international eyeball.

City intelligent power grid is to combine advanced remote gauging technology, information communication technology, analysis and decision-making technology, automatic control technology and energy power technology and highly aggregate with city power grid infrastructure to form a new modernized city power grid. City intelligent power grid can apply the best technological concept in the industrial world to the power grid to accelerate realization of city intelligent power grid such as open system structure, internet treaty, plug and play, shared technology standards, no specialized and interoperability and so forth. Some have been applied in the power grid but some are still in face of great technological challenges.

City intelligent power grid is a valid route for the whole society to realize energy conservation and emission reduction in a large range. Its construction not just brings about revolutionary changes to the power industry and will incur substantial revolution to people's daily life style and living habits. City intelligent power grid has the most significant effects such as profits resulted from reliability of power energy and updated power quality, profits from power equipment, individual and network safety so that intelligent power grid can continuously conduct self-monitoring, timely hunt for conditions that might endanger its reliability and individual and equipment safety and provide ample security guarantee to the system and its operation; profits from energy efficiency: efficiency of city intelligent power grid is higher. Guidance of interaction between end users and electric power companies for demand-sided management helps bring down demands for peak load and bring down total energy use and energy loss; profits of environmental protection and sustainable development: city intelligent power grid is 'green' and support of seamless connection of distributed renewable energy and encouraging of publicity and use of power-driven automobiles help reduce greenhouse gas, PM2.5 and emission of other toxic gases.

The development goal of city intelligent power grid is to build a city power network acquiring distributed dispatching management with intelligent judging and self-adjustment capacity, aggregate and use distributed power generation, notably renewable clean energy for power generation, apply ample imbedded intelligent equipment, distributed computing and communication technology to carry out actual-time monitoring, analysis and controlling on the operation status of city power network so that it can be equipped with the capacity of self-discovery and post-accident speedy recovery. By referring to two-way visibility, customers are encouraged, advocated and supported to participate in power market and provision of demand response in an effort to ensure high efficiency, high reliability, high power energy quality and power supply with reasonable price and to provide backup to such technologies as plug-in (hybrid) power auto, distributed photo-voltaic power generation, energy storage, energy-conservative towers and other technological applications.

Key construction content of city intelligent power grid incorporates:

(1) Advanced gauging system: by linking power companies with users, it contributes to cooperation and interaction between the two parties; big data it provides will produce huge profits. In the meantime, it will lay out the last section of two-way communication to the power grid so that the power grid can be gauged and opportunities can be provided to advancement of intelligent city.

(2) Distributed power connection: in the digitalized society in the future, requirements on security and reliability of power supply and quality of power are increasingly strict. Large-scaled connection to distributed power is an ideal plan and is a development bottleneck in the city intelligent power grid urgently in need of being redressed.

(3) Automation of advanced power distribution: automation of power distribution and advanced power distribution automation greatly contributes to reliability of power supply and improvement of asset use ratio. Its roles are multifold and it particularly shows significant effects when it comes to improvement of managerial level, managerial efficiency and user services.

(4) Advanced asset management: advanced asset management along with advanced power distribution automation and advanced gauging system can start from multiple aspects covering planning, building, operating, checking and maintenance of city power grid in an effort to realize use ratio of electric power devices and comprehensive use efficiency of energy.

(5) Electric car connection: as an entry point for building city intelligent power grid, electric car cannot just drive forth marketplace, boost public cognition on city intelligent power grid and become an acute weapon for load adjustment of city power grid as a distributed energy storage system.

Construction of city intelligent power grid calls for in-depth technological research in terms of grid structure, equipment level, communication technology and so forth. By combining with the status quo on city power grid both home and abroad, it conducts demand analysis and establishes a set of open standard systems characterizing strong applicability and sound compatibility. The government should fully give play to the guiding role, launch active polices on development of city intelligent power grid, align development strategies of city intelligent power grid to the national development and energy strategies, give out support on industrial policies in planning and building of city intelligent power grid and create a sound external environment for development of city intelligent power grid.

In construction of intelligent power grid in China, the following should be stressed:

(1) One characteristic of intelligent power grid is two-way mobility of power and information so as to build a highly automatic and widely-distributed energy swap network. To realize actual-time information exchange and realize quasi-simultaneous supply and demand balance on the equipment level, edges of distributed computation and communication are incorporated to the power grid. Our country is in particular need of boosting attention paid to development and use of distributed power (power generation, stored energy and demand response).

(2) The sequence of implementing intelligent power network has its values. Round unfolding is not economical and transition to completed intelligent power grid takes a long time. The structure of AMI communication system provides an opportunity for building of intelligent power grid and intelligent city. We should soundly put into effect top-level designing of communication grid from the national level.

(3) Core principles for intelligent power grid should take such into account as 'does the works we get engaged in apply to the market? Does it goad users? Does it realize asset optimization? Can it acquire efficient operation? So we should insist:

- Innovation-driven innovation to obtain myriads intellectual property, bring down cost of intelligent power grid and boost profits of intelligent power grid.
- Conduct sufficient analysis on cost profits in advance. Power Company and supervision outfits should continuously show to users that profits of intelligent power grid will eventually outnumber its cost and should ensure that reasonable and affordable power price is charged.
- Laws and regulations used to encourage power users, manufacturers and power companies to get involved in intelligent power grid need to be launched. Our country should quicken implementation of time-share/actual-time power price, open user-sided power market and vigorously facilitate R&D of 'plug and use'.

City Intelligent Energy Grid

In today's society marked by increasing paucity of energy resources and year-on-year rising of energy consumption, the energy system is an inevitable core round in city infrastructure. How to reasonable match the concept of intelligence and propose a set of intelligent resource system that can go hand in hand with building of intelligent city is an issue meriting our careful thought. The bulk of domestic cities are using traditional energy grid that does not sufficiently align with the concept of intelligent city proposed for the present and confronts the following challenges:

(1) At present, city energy grid is short of comprehensive top-level design and comprehensive coordination mechanism that can't balance clashes between energy (resource) supply and demand.
(2) A sound and extensive information network technology, mechanism and system are lacking and it can't conduct actual-time, efficient and handy response and transmission to supply end and user end of energy.
(3) It is short of valid managerial technologies and systems to tackle with massive data and falls short of satisfying accurate analysis and precise preplan to confront gigantic information in the intelligent energy network in a big data era.
(4) It lacks risk acknowledge and technology coping strategies towards building of intelligent energy network and can't form systematic energy grid building standards and profitability model.

Therefore, up against the aforementioned issues concerning city development, we have proposed the following concept of intelligent power grid as an energy grid characterizing diversified interaction, resource integration and optimized configuration with intelligent power grid, intelligent water network, intelligent fuel gas network, intelligent thermal power network, intelligent use of wasted resources, intelligent configuration of pollutant, intelligent emission control and inter-industry framework of energy and resource distribution as the basis, with multisource and stable power supply, clean and efficient energy use, speedy and handy transmission, secured and ample power reserve and performing emission and reduction as characteristics and information communication technology and intelligent data center as the reliance.

Core concept to build an intelligent energy grid is to coordinate and run varying energy resources, coordinate balance of different levels of water, power, gas, heat and coldness and attain efficient use of energy resources. It aims at realizing intelligent deployment of energy and resources with conservative consumption and environmental bearing capacity as comprehensive orientation, realizing mutual matching of supply and consumption of energy and resources, improving use ratio of energy resources and averting shortage or consumption of energy resources.

Intelligent power grid advocates that sustainable ecological environment be put as the premise, sustainable economic and social development be taken as the objectives and scientific and reasonable energy production means and consumption mode as well as corresponding incentive mechanism and constraint mechanism be established. Its connotation is mainly embodied in the following three aspects:

(1) Technological innovation: sustainable use of energy is taken as the goal, low energy consumption, low carbon and low pollution is taken as the basis and advanced new technology, technique and devices are applied to vigorously develop clean energy, actively develop renewable energy, new energy and distributed energy technology, optimize energy consumption structure, boost use efficiency of energy and bring down economic and environmental issues brought about by evolution of energy system.

(2) Institutional reform: put sustainable economic and social development as the goal, utilize advanced new technologies, techniques and devices, establish corresponding incentive mechanism and constrain mechanism, formulate laws and regulations related with energy development and use and eventually realize congruent development of economy, society and environment.

(3) Sustainable development: in the development and use process of energy, we should forever pay attention to protection of ecological environment, constantly work on development and use mode of energy and incentive and constrain mechanism, slow down restriction of energy bottleneck and pressure of ecological environment and bring to fruition sustainable development of human society and nature. Moreover, along with footsteps of urbanization, how to realize demographic urbanization in its true sense and make sure that agricultural population in cities assume their respective responsibilities is an issue in need of being resolved.

Development Strategies of City Intelligent Grid and Relevant Suggestions are Seen Below

1. Development objectives and intelligent network framework should be clarified

Building of intelligent power grid should be piecemeal but the overall design should not focus on fuel gas, power, thermal power, water affairs and such things and should make coordinated considerations. Some completed city intelligent energy system abroad is huge in input and prohibitive in price that hinders its promotion. Pertinent to the state condition in China, a reasonable energy grid framework befitting market demands is proposed. Intelligent power grid is developed with the aim at 'low carbon, high efficiency, hierarchical use, intelligent configuration and supplementary of advantages'.

2. Government should fully display the leading role

City intelligent power grid is a key measure to further promote industrialization and urbanization and is a major means to attain sustainable development. Government should give full play to its guiding role. Since building of city intelligent power grid is a complicated and systematic project that involves multiple branch network and overall platform designing, it requires that the government set up a committee that multiple departments take charge, make overall coordination on building matters and launch polices and suggestions for city intelligent power network.

3. Establish intelligent communities and form model effects soon

Building of city intelligent power grid calls for a development model with planes driven by spots. It requires formation of demonstration effect in the locality before promoting it to the whole city. Our country is yet to have a large-scaled pilot involving multisource systems in this field. We can draw reference from American or Japanese experience by establishing an entirely intelligent demonstration community before driving forth building of the whole city intelligent energy grid and boosting the industrialization level.

4. Reinforce building of basic information network and reform of existing branch network

Construction of a city intelligent energy grid calls for information communication platform as the backup and intelligent control as the means to realize high fusion of 'energy flow', 'information flow' and 'business flow'. Efficient, stable and secured information disposal technology is the core for intelligent energy grid building. Starting from the actual demands and following domestic and overseas experience, an open network equipped with secured safety strategies that is compatible with branch network of varying sectors can be built.

5. Quicken advancement of scaled development of new energy technology

Adjustment of existing energy structure and vigorous development of new energy industry and new energy-conservation technology is a major round for building of future intelligent energy grid. Popularization of new energy should, first of all, be reflected in planning by proposing predictable plan and making reasonable arrangement on planning of possible routes. Secondly, to reinforce research on reliability and safety of new energy technologies is also indispensable. So other than supporting new energy industry, government should sort out existing energy price system, reinforce popularity and promotion of various new model energy-conservative technologies, boost competitiveness of new energy industry and realize diversity of energy supply.

6. Guide companies and colleges to get involved in building of intelligent energy grid

As the main body of technological research, companies will undertake most research, manufacturing and building job in the building process of intelligent energy grid. Companies should form more close cooperation with schools of higher learning. With talents as the basis, company technologies and research and development project as the reliance and support of state policies as the guarantee, companies should boost capital input to building of intelligent energy grid and try to realize industrialization of city intelligent energy grid.

7. Intensify ethnical education in building of intelligent energy grid

Reliability of network and security of data will compose two major factors affecting whether intelligent energy can be accomplished. Once comprehensive or partial network collapse occurs disorder of city operation will incur. Thanks to interconnectivity of national confidential secret, private company information and individual private information, leakage of issues will bring about catastrophic consequences. So in framework and management of intelligent energy, legal environment is in urgent need of being adjusted in a bid to regulate users' behaviors in the information sharing process. Corresponding laws and regulations should be put on the agenda. Next, that information ethnical education should be boosted to facilitate residents to have building of information morality improved is also indispensable.

8. Strictly stick to implementation of people-centered policies and boost people's living quality

Urbanization is the only road leading to modernization, is the strategic key to transform development modes, adjust economic structure and expands domestic demands and a major route to address agricultural, rural and peasant issues, promote rural and urban coordinated development and upgrade people's living standards.

New technological products and services with promotion of intelligent energy grid are more adjacent to life and viably boost people's life quality.

9. Put into force industries and redress employment problems

Intelligent energy grid is a huge systematic project that combines artificial intelligence and various energy technologies and can redress energy issues at 'one package'. Promotion of intelligent power grid would inevitably bring about booming development of emerging sectors, put into force new energy sector, provide stable employment opportunities to surrounding residents before implementing household registration policies, addressing 'lingering' issues, fundamentally realizing the goal of demographic urbanization and contributing to practice of urbanization building.

Abstract of "Intelligent Manufacture and Design A Development Strategic Research in China"

General Concept and Background of Intelligent Manufacturing

Smarter Planet (Gan et al. 2009) brought up by IBM has raised great attention. Many places in China have started intelligent city construction. The concept of intelligent manufacturing is seen in the construction catalogues of some intelligent cities like Ningbo (Tong 2010), Shunde (Xu 2012), Fuyang (Fang 2011), Huzhou (Shi and Tan 2012) and etc. The intelligent city construction of these cities includes intelligent manufacturing and design, intelligent industry and the contents of intelligent manufacturing.

Intelligent manufacturing is a new industrial form after the deep integration of informatization and industrialization. It emphasizes the adoption of "intelligent" technology to integrate and optimize the links of manufacturing enterprises, such as design, production, managing, service, business and etc. Thus the overall competitiveness of manufacturing enterprises will be improved and the development tendency of manufacturing skill from mechanization, automation and digitization to intellectualization will be reflected.

The Eighteenth National Congress points out that "We should keep to the Chinese-style path of carrying out industrialization in a new way and advancing IT application, urbanization and agricultural modernization. We should promote integration of IT application and industrialization, interaction between industrialization and urbanization, and coordination between urbanization and agricultural modernization, thus promoting harmonized development of industrialization, IT application, urbanization and agricultural modernization." Intelligent manufacturing is a manufacturing mode which deeply integrates informatization and industrialization.

Intelligent manufacturing system is a human-machine integrative system. It enlarges, extends and partially replaces the mental work of human experts during manufacturing, and even realizes unmanned operation in partial operating range, which improves manufacturing level and production efficiency. Intelligent manufacturing is the outcome of a dual function of technical drive and demand pull. Driven by the new generation information technology and under the pressure of constructing a market of individualization, greenization, high end, and globalization, intelligent manufacturing develops from a top-down and centralized mode to a bottom-up and decentralized mode (European Commission 2004).

America has great expectation on intelligent manufacturing, hoping to realize the return of manufacturing and improve its competitiveness and employment rate through intelligent manufacturing. In the past, due to the competition of low-wage countries, especially the rise of China, American manufacturing outflow caused depression and bankruptcy in many traditional manufacturing cities. Western scholars think that the generation of the third industrial revolution represented by intelligent manufacturing may end "China's rise".

As one of the strategies in "High-tech Strategy 2020" brought out by German government in November 2011, Industry 4.0 is considered the fourth industrial revolution in Germany, which aims at supporting the research and renovation of the new generation revolutionary technology in industrial circle and keeping Germany's international competitiveness. Its essence is to construct a highly flexible intelligent manufacturing system of individualization and digitization. Industry 4.0 will bring a brand new change in working mode and environment. New collaborative working mode enables work to separate from factories and to develop through a virtual and mobile way. The employers will enjoy high autonomy in management so that they can actively plunge into and regulate their work (Luo 2014).

Impact of Intelligent Manufacturing on Intelligent City

Intelligent city is in great need of intelligent manufacturing. And intelligent manufacturing is the main embodiment of the innovation capacity of an intelligent industrial city, a necessary method to improve its ecological environment, and an essential infrastructure in an intelligent city. Intelligent manufacturing helps to keep citizen's basic necessities of life healthy and safe, provides fine-quality manufacturing services, supports structural layout optimization, and promotes employee's happiness.

Intelligent manufacturing limits cities from endless expansion, enables people from outlying areas to participate into the commercial tide, and let them have more chance to contact with nature instead of waiting in traffic jams.

Cities in different areas have different forms. Figure A.1 depicts different intelligent manufacturing needs of cities with different forms. Most of the cities in China are processing and manufacturing cities. Undoubtedly, intelligent manufacturing is significant to the economic and social development of manufacturing cities.

Therefore, intelligent manufacturing should be the foundation and one main concern in the construction of intelligent cities, especially intelligent industrial cities.

Different with the intelligent buildings, transportation, medical care and safety guarantee systems in intelligent cities, intelligent manufacturing has a long-term, large-scale and vague impact on intelligent cities. For example, innovation capacity needs long-term fostering and it has long-term effect on city development. For another example, the infrastructure, the guarantee of healthy and safe basic life necessities and the fine-quality manufacturing services provided by the development of intelligent manufacturing are large-scale, not just limited to the local places. Intelligent manufacturing can help to promote intelligent industrial city employee's

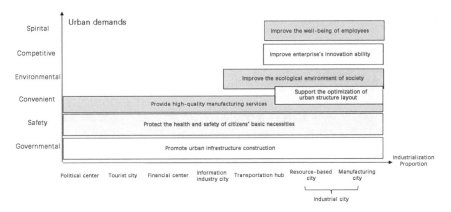

Fig. A.1 Different intelligent manufacturing needs of cities with different forms

happiness, which is hard to determine quantitatively and has long-term effect. Undoubtedly, under the present government performance appraisal system, these long-term, large-scale and vague effects limit the enthusiasm of intelligent manufacturing project setting in intelligent city construction. We need to give a further consideration of how to integrate intelligent manufacturing and intelligent city construction more tightly.

The Main Content and Development Direction of Intelligent Manufacturing

Intelligent manufacturing includes intelligent design, intelligent processing and intelligent service as shown in Fig. A.2.

1. Intelligent Design

Intelligent design enables enterprises to develop new products and enter the so-called "blue ocean", which can gain high profits. It can also help to quickly design diversified and individualized products to meet the market demands so as to make big profits. And its design of low-cost and low-energy products can help customers save cost as well as enterprises and the society gain profits.

The supporting tools of intelligent design includes:

(1) Professional intelligent design software

Each product has its own characteristics and domain knowledge General design software only basically comes to use in the latter period of product design and cannot actually solve the problem of product innovation. Therefore, many big foreign enterprises have their own special softwares and models which are the result of their long-term research and also their core competitive capacity. This software

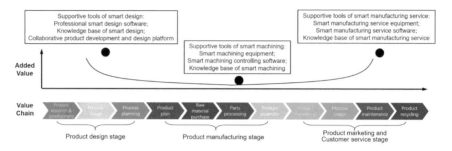

Fig. A.2 The content and impact of intelligent manufacturing

coheres and solidifies the structural knowledge obtained in the process of a long-term product development and design, which is strongly targeted and intelligent. It enables enterprises to have massive designs based on knowledge and simulation, improves the ability of computer aided design, and particularly lets young designers get into their roles quickly. Generally, this software is not available to buy. Even if you get it, it'll be difficult to use, because only people with excellent professional background can build applicable models and use the system properly.

(2) Knowledge base of intelligent design

In professional intelligent design software, knowledge can be transformed into program and knowledge base controlled by inference machine. While the knowledge in intelligent design knowledge base is disordered and constantly updated explicit knowledge which comes from many different disciplines. This explicit knowledge becomes the implicit knowledge in expert's head by expert's learning, which can improve expert's design capacity. And finally, experts will do innovative designs. Because lots of knowledge cannot be transformed into design program in a short time. The intelligent design knowledge base involves many different kinds of knowledge which is quite a mixed bag, so we need to screen them. Employees need to release new knowledge frequently and evaluate the value and relationship of the knowledge.

(3) Collaborative product development and design platform

Product development and design involve many disciplines and need the collaboration of many employees both within and outside the enterprise. The collaborative product development and design platform facilitate this collaboration. It can help to find employees with some specific knowledge quickly. We can carry out collaborative design between enterprise and customers. After knowing customer's demands, we can even let customers design by themselves. We can carry out collaborative design among enterprises, partners and upstream and downstream firms of product value chains. The design knowledge involved here is mainly implicit knowledge.

2. **Intelligent Processing**

Product intelligent processing is also called intelligent manufacturing. In order not to mix the concept of manufacturing in a broad meaning and narrow meaning, we use the term "intelligent processing" here.

Intelligent processing is basically the generalization of processing method, which can substitute the technical work workers used to use.

The applied range of product intelligent processing mainly includes:

(1) processing sophisticated products and assembly parts that cannot be processed by workers;
(2) processing sophisticated products and assembly parts under harsh conditions
(3) rapidly making diversified and individualized products to meet market demands and gain big profits for the enterprise
(4) processing products with low cost, low energy consumption and other resource consumption, so as to help customers save cost as well as enterprises and the society gain profits.

The supporting tools of intelligent processing includes:

(1) intelligent processing equipment (also intelligent manufacturing equipment)

Manufacturing equipment with perceptive, decision-making and executive functions are collectively known as intelligent manufacturing equipment including intelligent assembly line, intelligent manufacturing units, intelligent processing center, intelligent industrial robot and etc. Intelligent processing equipment can enlarge, extend and partially substitute human expert's brain work during processing, and even realize automatization within part of the working range so that the processing level and production efficiency can be improved; realize unmanned intelligent processing under hostile environment (like high concentrated hazardous substance, intense radiation, high temperature and etc.) so as to lighten the negative impact on employee's health; process some oversize products and sophisticated and tiny products, because these product's processing is beyond manual work capability; realize a high efficient and high flexible intelligent processing.

(2) intelligent processing control software

Control software is one core of intelligent processing, which coheres and solidifies expert's processing experience and realizes automated processing.

(3) intelligent processing knowledge base

During the processing, large amount of knowledge will be needed to support employees to conduct intelligent processing. Intelligent processing knowledge base resembles intelligent design knowledge base.

3. Intelligent manufacturing service

Through intelligent manufacturing service, manufacturing enterprises efficiently extend following services:

(1) Highly individualized service: individualized service not only saves resource and time, but more importantly, it can provide specific solutions for individual situation so as to improve the chance of solving problems.
(2) Product leasing shared service: for example, automobile leasing and renting can reduce the use of cars and resource waste. Intelligent service reflects in: when customer needs to rent car, the system will quickly get the customer's data including the customer's most familiar vehicle model, driving habits and etc. and deploy car from the nearest place; when the renting car is in use, the system masters real-time conditions of the car, providing the customer safety guarantee services; when the customer returns the car, the system will immediately send the nearest service staff to fetch the car. It is estimated in Japan that its electric automobile shared service revenue will reach 150 billion yen by 2020.
(3) Product re-manufacturing service: such as automobile engine re manufacturing service, re manufactured engine has the same performance and service life as new engine, with half the price of the latter and less than 10% resource consumption of the latter.
(4) Energy-saving service of energy-using products: through methods like intelligent remote monitoring, monitor energy consumption of energy-using products (like air-conditioner, refrigerator and etc.) and send serviceman as long as the product energy consumption reaches a certain level so that the product keeps a low-energy state.
(5) Product value-added service: enables customers to enjoy new services with the original product through changing modules of product and providing new software, such as the new services for mobile phone users: automobile navigation service, price comparison by taking photos of the products and etc.
(6) Product life-cycle management service: provides management services in each step of the product lifecycle, like maintenance service, recycling service and etc.

Through providing fine-quality manufacturing services for intelligent city, the manufacturing enterprises reach the following goals based on intelligent manufacturing:

(1) Promote the development of manufacturing service and increase green GDP, because resource and energy consumption in service is less than that in product manufacturing.
(2) Manufacturing enterprises extend to service and increase revenue through service so as to improve competitive capacity.
(3) Through service, manufacturing enterprises improve customer's satisfaction and understand customer's needs so as to improve product innovation capacity.

Developed service industry is related to develop manufacturing industry. Traditional manufacturing in the low end of value chain cannot lead to a developed service industry, unless the city has special natural and cultural landscape and its development counts on tourism.

Figure A.3 depicts the concept of intelligent manufacturing providing fine-quality manufacturing service for intelligent city.

Through intelligent manufacturing service, we can create many jobs of low resource consumption and low environment pollution, extend product lifecycle, or keep energy-using product in a low energy consumption level during its lifecycle. This can not only save cost for customers, but also can gain profits for enterprises and the society.

The main supporting tools for intelligent manufacturing service are:

(1) Equipment and appliance for intelligent manufacturing service

Manufacturing service state/environment intelligence and sensing equipment and appliance can help enterprises to know the operation condition of customer's products and conduct remote monitoring; can provide service for customers with service robots; can re-manufacture products with re-manufacturing equipment so as to extend product's service life; can provide overall solution service.

(2) Intelligent manufacturing service software

Help enterprises provide online long-distance "one-to-one" product using and maintenance service; provide many new professional value-added services;

(3) Intelligent manufacturing service knowledge base

In manufacturing service, a large amount of knowledge will be needed to support employees to conduct intelligent manufacturing. Intelligent manufacturing service knowledge base resembles intelligent design knowledge base.

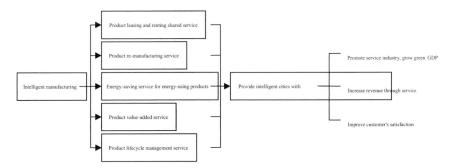

Fig. A.3 Intelligent manufacturing providing fine-quality manufacturing service for intelligent city

Abstract of 'Information Environment Construction and Big Data of Intelligent City A Development Strategic Research in China'

As China's urbanization picking up speed, both emerging cities and developed old cities have been given unprecedented economic, political and cultural functions, and inevitably pushed to the center stage of China's economic and social development, in which they play a leading role. Meanwhile, along with city development, there emerged many severe problems like excessive consumption of resource and energy, environment pollution, traffic jams, housing shortage and etc. There are also challenges like employment, medical care, education and etc. Therefore, promoting China's city intelligent development in a scientific, conservative and green way has been an important and urgent task.

New Understanding of Intelligent City

Human's pursuit for smart city has a history over 2000 years. Whether it's Plato's Utopia, Howard's Garden City, or Le Corbusier's Urbanism, they are all filled with Utopian characteristic and unlikely to solve big city malaise today. In Plato's Utopia, it's intelligent, well-organized and controllable. He thinks that only philosophers can be the leaders of the Utopia, because instead of chasing money and fortune, they pursue intelligence. Today, we have found the path to intelligent city construction, which is the information path based on big data and cloud computing. Through the link of big data, human society (including philosophers), physical world and information space merge together and become an intelligent Ternary world.

In 2008, IBM brought up the concept of "intelligent city". Since then, many cities around the world have implemented intelligent city development plan and consider it the trend of future city development. The latest report released by U.S National Intelligence Council says: from now to 2030, there will be 13 important technologies influencing the world, one of which is "intelligent city" technology. In recent years, over two hundred cities in China have queued up to put forward intelligent city construction project, which plays a positive role in China's economic and social development. However, without knowing clearly the development periodicity and uniqueness of China and its cities, we performed, to some extent, blindly in "spare no effort to get on with the plans", which caused some after effect.

China's city development is clearly in a different stage with developed countries. Developed countries' city development has gone through the era of large-scale urbanization and industrialization. They are now mainly in the stage of intelligent city management and service. And China is now in the stage of large-scale urbanization construction, facing difficulties and problems that are different from foreign countries. The path of China's city development is certainly unique. We are now in a stage with the integration of informatization, industrialization and urbanization. Mere consideration of city management and service intelligentization is far from enough. We should also give comprehensive consideration to national conditions and demands relating to China's city development, like technology, policy, law, culture and etc. To construct intelligent city, we need not only advanced information technology to build "the third platform" represented by cloud computing, but also need to construct a people-oriented and impartial justice system and a harmonious culture atmosphere of collaboration.

During the consulting research of this project, we have found some new understandings of intelligent city:

1. intelligent city emphasizes on meta-synthesis based on big data and integration knowledge
2. intelligent city (integration knowledge) = $\sum$ intelligent city (domain knowledge) + big data technology

"Intelligent city" construction implemented in China now is basically the processing of information resource under many applied discrete frameworks and a thinking model of splitting the problem and solving it part by part. It emphasizes on solving one specific problem. The point is "splitting the problem", and then "solve the split parts". The "intelligent city" advocated by Chinese Academy of Engineering emphasizes on meta-synthesis integration knowledge. The past smart city construction provides knowledge accumulation for the integration of intelligent city. Intelligent city is the treasure chest of big data. Data technique is the scalpel for curing big city malaise. Problems in intelligent city construction can be solved by big data technique. In fact, intelligent city construction is a process of developing a city rapidly, soundly and economically. We should deeply integrate China's urbanization 2.0, informatization 2.0 and industrialization 4.0, and lead a path with Chinese characteristics.

Intelligent city is a new stage of city construction with "four modernizations" led by informatization under the new normal economy.

Based on digitization and network, intelligent city is a new stage where city develops from informatization to a high-end level (intelligentization). Cloud computing and big data is the key technology to realize intelligent city and also the new lever to increase productivity under the new normal economy development, playing a leading and driving role in economic transition. New trends like internet thinking, combinatorial innovation, maker and DIY are all related to cloud computing and big data. The intelligence source of a city comes from the full use of computer and internet, from the decision based on data, information and knowledge.

Intelligent city construction should lead an application-driven path instead of a technology-driven path.

Intelligentization is one of the connotations of a city's scientific development, but not an ultimate pursuit above everything else. Intelligentization is an everlasting dynamic target. Both technology and demand are developing. "Intelligent city" will aim at another higher target over a period of time. The level of a city's intelligentization is related to basic conditions of informatization. We should not try to accelerate blindly. The aim of developing information technology is to serve people. And the sole criterion for testing all technologies is application. Technology may be limited, but application is limitless. We should adhere to the development strategy of "application first" and a technical route led by application. Instead of leading a technology-driven route, we should set realistic targets based on practical situation, keep up with times and adjust measures to local conditions. To build an intelligent information environment, we should arouse the enthusiasm of application departments and innovating enterprises through policies and different measures, and develop new applications through crossover combinatorial innovation, finding a way out from applications.

Build a Favorable Information Environment Before Intelligent City Construction

There are five connotations in China's intelligent city development: (1) in basic environment construction, we should construct basic information infrastructures including perception terminal, information network and cloud computing center, comprehensively supporting information communication, service delivery and business collaboration among the city public, enterprises and government. (2) In people's livelihood service, we should improve citizen's awareness and capability in applying information technology and let them have easy access to different information service so as to improve their living quality. (3) In industrial development, we should raise innovation energy and city competitiveness with the help of information technology and realize sustainable development. (4) In government management, we should achieve high efficiency and accuracy in city governance and improve management efficiency and service effect. (5) In system mechanism and culture construction, we should give full play to scientific decision-making mechanism based on data, and build a culture system with intelligent city characteristics.

To achieve the targets above, the top priority is to realize information environment construction in intelligent city. Information environment is the sum total of the elements related to information activities. It involves the consideration of the impact of information transmission and application on social development, and also the impact of many social elements on information activities. At a macro level, information environment includes social (cultural) environment, economic environment and technological environment. Constructing intelligent city is essentially constructing a working and living environment that meets the needs in information era. It not only involves the hard equipment environment of information acquisition, transmission, saving and processing, but more importantly, the soft environment which reflects people's cognition and inner demand.

The social environment of intelligent city reflects government, enterprises and citizen's understanding of and demand for information era. If the local government and citizens don't have proper awareness of information era and still stay in the industrial stage where a lot of non-renewable resources are consumed, then they cannot conduct intelligent city construction attentively. Therefore, the premise of constructing intelligent city is informatized education and conducting information consumption demand. Policy environment is an important aspect of information environment. The requirement of constructing intelligent city is to make an advisable policy in which the industrialization, urbanization and rural modernization are led and driven by informatization.

Economic environment is often taken as the ecological environment of information industry, adopting ecological theory and method to investigate the living and developing environment of enterprises. For intelligent city construction, the particularly important economic environment is small and medium-sized innovating enterprises and open source communities. Only through incremental innovation of many small and medium-sized enterprises can various data opened by government form information service on a large scale. Many network service providers in China, such as Alibaba, Tencen, Baidu, has entered the top ten in the world. These leading enterprises are the significant force in intelligent city construction. But if the policies toward intelligent city construction only favor the leading enterprises today, then there might emerge monopoly, and innovation would be stifled. The most efficient way to develop new technology like big data is through open source community and crowd sourcing. China's contribution to open source community doesn't match its economic status, to which we should pay high attention.

The technological environment of intelligent city includes elements like port, net, cloud and etc. Intelligentized Internet of Things (sensor network) forms the "nerve terminal" of intelligent city, interconnected and interworking information network form the "neutral network", the business application platform based on cloud computing and big data analytic forms "nerve center system", and the massive city data form "blood and nutriment".

Broadband and the wireless network is the most fundamental technological environment of intelligent city. The BBN (Broadband network) construction in China lags far behind foreign countries, which leads to a continuous falling in the ranking of some informatization indexes. (In 2014, China's Networked Readiness Index fell by four rankings to No. 62nd) It is reported that we have made great progress in recent two years. It is estimated that by 2015 the FTTH users will top 80 million and broadband users will get to 40%, which is above the average of developed countries. China's broadband is only above 4 megabytes, while other countries in the world have more than 20 megabytes broadband. The actual feeling of broadband users in China is less satisfied than foreign users. So instead of being unrealistically optimistic, we still have a long way to go in network construction.

The information environment nowadays is not a human-computer symbiosis system of man-computer, but a ternary-world integration system of "man-computer-object". Logically, the future information environment of intelligent city would be integrative. To build the information environment of intelligent city, we should overcome the drawbacks of "information islet, information chimney" in connectivity, and avoid "fragmentation" of city information environment.

MIIT, MOST, MHUD and Tourism Administration have respectively implemented intelligent city plan. China's intelligent city construction still lacks integrative deployment of information environment construction.

Information environment is in a continuously optimizing and evoluting process. The information system we build now might not suit the new requirement and framework of big data technology development, so it needs ongoing reform. We should keep a perspective view on the evolution of information environment. Intelligent city construction costs high. It costs 35 billion dollars to construct a 6 sq km intelligent city in Songdo, South Korea. We should have cost consciousness in the information construction of intelligent city, being realistic and conducting in steps. We'd better not put too much into the information technology that still doesn't have mature standards. In the years to come, we should promote vigorously the construction of standards and norms relating to intelligent city and give high attention to policy making and public education.

Challenges to City Information Environment Construction Brought by Big Data

Big data is not just a tool, but also a strategy, world view and culture. It will bring a social reform, especially the reform in public management and public service. It will also bring new challenge to city information environment construction. In big data era, enterprises turn their focus to data. Computer industry is transforming into a real information industry, from the pursuit of computation speed to big data processing ability. Software industry will transfer its priority from programming to data. The rising of big data processing system changes the developing direction of cloud computing and make it enter the Cloud 2.0 era symbolized by Analysis as a Service (AaaS).

Big data technology originates from internet industry. And the most successful application so far is also in internet industry. Chinese government agencies at all levels and traditional industry have accumulated massive data during daily management and business operation. But the big data application in these industries is still in a primary stage. The real value of big data reflects in the application in different industries and people-benefit service. The problem needs to be solved immediately in intelligent city construction is how to wake these sleepy big data resources, achieve scientific decision making in management, and open up new business mode.

The big data analytic technique is now mastered by the minority of internet enterprises and scientific research institutions. But they either possess only the data (like internet industry) of their own industry, or lack of the sample data used to verify techniques. However, the institutions truly possess the data, like government agencies and traditional industries, lack necessary big data analytic techniques. The major concern of city information environment construction is how to get out of this dilemma, open up industry chain and achieve the unification of technology and resource.

"Big data" is massive, rapidly growing and diversifies information asset. Big data technology is the kind of technology with which we can acquire valuable information quickly from different kinds of big data. The generally referred "big data" is not only the data itself, but also includes tools for collecting data, platform and data analysis system. The challenge brought by the big data era doesn't only reflect in how to acquire valuable information and knowledge from big data, but also in how to improve and develop new tools and systems, how to collect, transmit, save and process big data more efficiently, and how to construct more intelligent and energy-saving big data analysis system.

For big data analysis, the most important infrastructure is storage device. As the rapid growth in data size, storage device should achieve high extendibility flexibly. Big data analysis involves the tracking toward social media and transaction data. In order to conduct real-time decision-making, the storage system delay must narrow down to millisecond level or even microsecond level. Storage device should be able to process data from different origin systems at the same time. New-type storage devices like SSD have gained more and more attention gradually.

Due to the surging number and diversified data types, the present data base technology and data analysis doesn't suit multiplex mode mixture data and the requirement of quick and real-time processing. Some hypotheses that used to seem correct are now invalid in big data application. New technologies like distributed file system, distributed data base, stream processing and figure computing have become the hotspot in data management and analysis. Collaborative design of software and hardware and "software define" have become an irresistible trend. Essentially speaking, "software define" aims at separating integrative hardware device into several parts and build a virtualized software layer for these basepieces.

Big data safety also faces great challenge. Improper processing of big data can cause great harm to users' privacy. The threat people facing is not only confined to personal privacy disclosure, but also the prediction of people's state and behavior based on big data. Even anonymous data could be identified with "de-anonymity" technology. Data saved in big data system doesn't enjoy the same level of protection as traditional data base does. The application of technologies like digital signature and message authentication code to the authenticity detection of big data faces huge difficulty. False or wrong data could lead to safety disaster.

In the ecosystem of big data and cloud computing, there already exists widely-adopted open source software such as Hadoop, Spark, Openstack and etc. Among nearly ten thousand core community volunteers around the world, there are less than 200 in China. During the 13th Five-year Plan, we should endeavor to cultivate 1000–5000 core community volunteers, making more contribution to open source software. Independent innovation doesn't mean developing an existed software by ourselves, but means leading the world and participating in global development, application and maintenance. We should endeavor to construct independent and cooperative ecosystem, develop two or three open software systems led by China and engaged globally, and get rid of the constraint of closed proprietary software.

The Three Important Issues in the Construction of Intelligent City and Big Data Information Environment

1. Principles and realization approaches in city data continuity management

Data continuity management consists three parts: (1) digit generating file: saving with digits and reuse sustainably. Therefore, people who form, save and use the files should build a file management mechanism that followed by everyone, and maintain cooperatively the authenticity, reliability, integrity and availability of the electronic file during its lifecycle. (2) Digit generating information: maintaining with digits and ensuring availability, dependability and sustainable reusability. Thus, people who collect, save and serve the digital information should build dependable digital information cross-system and information managing framework with interconnection, intercommunication and inter-accreditation platform, maintaining collaboratively the quality of digital information. (3) Digit generating resource: managing with digits and ensuring trackability, traceability, associability and controllability. Trackability refers to digital resource being able to predict and simulate its evolutionary trend in chronological order, which can be applied to public opinion analysis, simulation test, market prediction and etc; traceability means digital resource being able to trace back to its historical version, which can be applied to evidence chains discovery and reliability assessing of digital contents; associability refers to digital resource being able to open up relevance and cross-domain access, avoiding fragmentation; controllability refers to digital resource being able to get risk control and protect personal privacy safety and national information safety. If digit continuity is out of control, there might be information loss, memory loss, unidentified identity and the disclosure of personal privacy and national secret. Digit continuity strategy should be included in the category of intelligent city so as to improve China's modernization ability of national governance.

In order to ensure accessibility, reliability and sustainable reusability, trackability, traceability, associability and controllability, and promote data quality and digital management ability, countries like Britain, America, New Zealand and Australia have made a series of digit continuity strategies.

There are a very few institutions of China's government departments archiving documents in digit ways. Most of them save electronic files in paper photocopies and then make them digital files for use with electron scanning. Saving with both paper photocopies and electronic files causes repeated work and the waste of human, financial and material resources. Research finds that due to the lack of state-level digit continuity strategy plan and top-level design, the government information resource is incapable of storing, crediting, using and controlling. We have done research and analysis on sixty-three regulations and standards relating to data resource management and use. We have also conducted on-site inspection in ten cities on twenty-eight data centers relating to the collection, custody and service of data resource of government administration. The result shows: none of the relating files covers the whole lifecycle, complete flow and total management factors of data resource, lacking unified quality standard for data collection, lacking legitimated electronic certificates, lacking a rule of sharing and exchanging unified data across departments, systems and regions, lacking methods of opening up and using data and rules of personal information protection; lacking the continuity management plan for information with whole lifecycle, complete flow, total factor and total quality, which considers government administration data resource as city public asset, lacking countermeasures for sustainable data reuse, such as digital certificates, digital memory, digital identity authentication.

Focusing on the existing problems, the research group puts forward the basic principles and realization approaches of China's city data continuity management from three levels: (1) following subject alliance principle at macro-level, making action plan for intelligent city data continuity, implementing data resource assets management, and constructing management system, verification system and authentication system for data assets register; (2) following activity interconnection principle at meso-level, making standardized management standards for intelligent city data processing activities, implementing data resource risk management, building the framework of authorizing and permitting an open access to data and the agreement management system; (3) following factor connection principle at micro-level, implementing automated management of electronic documents and online authorization and permission of open access to data, connecting automated management of data documents with network-based management of digital information and intelligent management of digital resource, and implanting them into city information infrastructure.

Therefore, it is advised that data resource continuity management and the planning of sustainable use should be added into national development planning for electronic government administration during the "13th Five-year Plan".

2. Open access to government administration information resource

Government data is a big fortune and also the key to an intelligent government. Intelligent government is closely related to government open data which forms an ecosphere pushing vigorously forward a fast and sound development of intelligent government. Making and implementing data open access should be added into the declaring and approving process of intelligent city construction project and the performance appraisal of informatization project.

In 2011 Brazil and other seven countries jointly released "Open Data Announcement", in 2013France, America, Russia and other countries of G8 signed "Open Data Charter", in 2014, 63 countries including India, Brazil, Argentina, Ghana, Kenya and other developing countries, built open partnership among governments, set web portals access to data. Open access to government data has been an international trend. According to the global development data index in 2014 released by Open Knowledge Foundation, China ranks 63th (in 2013, China ranked 34th). The situation of China's government data accessibility doesn't only lag behind developed countries, but also many developing countries.

According to the revelation and investigation on the laws and regulations relating to the openness and utilization of domestic government administration information resource, the problems need to be solved are as follow: (1) there is no laws or regulations basis for whether or not the government information accessed by individuals and enterprises can be used in business development; (2) lacking the online licensing mechanism of open data, data are only publisized but not accessible; (3) government data websites are deficiency in contents, single in types and uneven in data resource quality; (4) digital files have no legal certificate effect, lacks capitalization management system and interoperate regulation of data resource, and etc.

In this research project, the opening and utilizing of government administration data should abide by the following principles: (1) openness principle: opening and sharing government data with legitimacy, compliance of regulations and conformance of standards, unavailable openness and utilization is exceptional; (2) security assurance principle: ascertaining the range, risk grade and permission of data openness and utilization according to security level; (3) value-added orientation principle: promoting socialized value adding openness and utilization of data resource, emphasizing on the commonweal and commercial utilization of data resource; (4) quality guaranteeing principle: complete and credible contents, user-friendly data format, real-time updating. (5) duty-power-benefit principle: government taking charge of the quality of data collection, authorization of opening and using of data, and the service platform of data resource, data developers and users taking charge of data using behaviors after downloading. (6) Digit continuity principle: life cycle management on government data, collecting legally, registering compliantly with regulations, updating conformably with standards, storing safely and processing timely, keeping the continuity, consistency and standardability of management activities.

Introducing market mechanism is an important approach in promoting government data openness. We can integrate data resource formally with an API accessing way. And the management and maintenance of data set are still implemented by present government. Method of API accessing enables real-time charging. Government open data can also be collected, refined and classified by the third party portal, generating usable data set for the public and enterprises.

There are high hopes in data opening, but it won't reach its mature period in 5–10 years. Government data opening is a project with great funding and time and manpower consumption. We should analyze calmly: (1) Is the timing mature for the large-scale opening of government data? (2) Which data should be opened first? (3) How much social and economic benefit can data opening bring? Considering these problems, members of this research project assisted the government of Dongguan city to conduct a thorough real-name investigation.

This investigation is conducted based on the web data from the opinion collection platform of Dongguan government open data and the papercopy of "Opinion Collection Questionnaire of Government Open Data". Web data includes overall page view, differentiable visitor number and etc. 5000 paper questionnaires have been given out and 4264 valid questionnaires were reclaimed (the efficient reclaim rate is 85.28%). The public data directory (data of unopened directory) of Dongguan government includes seven categories which are cultural education (32), environment (25), people's livelihood (10), economy (6), agriculture (4), safety (3), medicare (2), with a total number of 82.

The investigation and statistic analysis shows that: interviewees support the data opening for most public directory. Data the interviewees would most like to see opening is the environmental data (ranking in the top four most favorable) relating closely to themselves, and the next is the cultural education data. The openness of data that obviously involve personal information meets strong opposition (the opposition rate of "Dongguan student athletes information" reaches 19.58%). And negative information data (directory names including key words of "punishment", "unlicensed" and etc.) will also be opposed by the public, which, comparing to the data opening of personal information, stands much less opposition. There is a positive correlation between educational background identity and support degree. Interviewees with Bachelor's degree are positive about government opening data. People with lower education degree show higher unconcern rate. College students and white-collar workers are more positive about government opening public directory data than other occupations. People waiting for employment, domesticated housewives and farmers show lower unconcern rate than any other occupations. Interviewees in business are rather discreet about government opening of public directory data.

The visitors on opinion collection platform of Dongguan government opening data are only 4028 from July 1st when it was launched to December 16th in 2014. Differential visitors are only 564, with few page visits and high jump rate of 54.26%. It indicates that at present the public don't know much about government opening data, let alone their requirements. Over the corresponding period, the most visited "basic information of real estate development company" on Shanghai government data service website only got 3147 page views, and the most downloaded "basic information of local police station" only got downloaded 4740 times. Comparing with the large population in Shanghai, there is little attention on government data website.

The research result shows that: at present stage, government doesn't need the urgent to open massive data to the public. In the coming two or three years, we should promptly implement legislation and make open regulations, popularize and improve people's understanding of government opening data, and make preparations for the large-scale openness of government data. At present, we can firstly open the environmental data and cultural education data that people care the most, and for enterprises, we can firstly open data of business administration, human resource, government's support for enterprise technology, and etc.

3. Management of government administration information resource and protection of personal information

There are "two wings" and "two wheels" in intelligent city construction: "two wings" is the opening and sharing of government data, "two wheels" is the protection of personal privacy information. We should make laws and regulations on these two aspects, which are in favor of healthy development of intelligent city "data opening" and "information security" are two sides of one object. We need to stick to the principle of "promoting security with development and guaranteeing development with security". It's not plausible to guarantee security with no cybers, no sharing, no interconnection and no intercommunication.

The development trends of personal information protection in foreign countries are as follows: (1) from inner-institute protection to cross-institute cooperative protection, from public sector protection to the protection among public sector, private sector and the third party cooperation agreement; (2) promoting simultaneously technical measure, administrative procedure and law compliance,; technologies like data masking preventing data from revealing, unidentifiable processing of personal information, and etc. (3) integrative management of full life cycle, complete flow and total factor. The following is the referential experience of personal information protection we can take from foreign countries: (1) from the scientific perspectives of law, management, sociology and information technology, considering the requirements of information technology for private data masking processing, security guaranteeing and risk control, considering various influencing factors and their interaction. (2) making multi-stakeholders agreement, ascertaining the power, obligation and duty of all interested parties including administrator of personal information, information agent, producer of information products, servers, customers and etc. (3) abiding by the principle of "joint while different", classifying personal information into general information and personal sensitive information; based on the specification of general information management process, making more strict requirements for personal sensitive information management. There are three new characteristics of personal information protection under the background of big data: (1) personal information category and value increasing; (2) diversified supplying and utilizing environment for personal information; (3) risk increasing through personal information processing

Up till now, there is no specific information law for personal information protection, and no personal information life-cycle management policies and regulations that applied to national institutions. The existing laws mainly involve financial insurance, less concerning government administration; the existing regulations about network environment mainly target on web service providers, teleservice operators, internet information service providers, not suitable for national institutions; the existing laws give most attention to protection, lacking the perspective of protection and utilization balance. The requirements involving personal information protection in China's laws focus on security and confidentiality, scare concern on legal and secure utilization.

Combining international experience and national status, this research project brings up seven principles of making countermeasures for personal information protection and information resource development and balance. (1) legal and convenient for people principle; (2) security maintenance principle; (3) whole process management principle; (4) informatization management principle; (5) risk orientation principle; (6) cooperative management principle; (7) integrative management principle.

Abstract of 'Information Network of Intelligent City A Development Strategic Research in China'

Intelligent city is a new period in city development. For a sustainable city development, more efficient operation of economic society and a better social livelihood, it aims at solving problems and challenges in city construction and management, industry upgrading and social livelihood with a new generation of information communication technology.

information network technology provides important support for intelligent city construction, and meanwhile, intelligent city creates fine development environment in favor of innovation for the research, development and application of information network technology. With mutual promotion, this will bring subversive change for people's production and living.

At present, there are many new developing trends in information network technology represented by broadband network, mobile internet, internet of things, cloud computing, future networks and etc. Taking intelligent city construction as an opportunity, China should fasten the development of new generation information network from aspects of core technology, marketing and industry chain construction, providing powerful support for intelligent city construction. The construction of intelligent city information network should closely surround the requirements of application fields like intelligent transportation, intelligent medicare and intelligent power grid, and build horizontal city information network infrastructures with the characteristics of ubiquitous intelligence, opening and sharing, heterogeneous integration, seamless mobility, security and credibility, green and energy saving, simpleness and transparency, flexibility and extensiveness, and etc.

The Background of Intelligent City Information Network Construction

1. Overall strategy of intelligent city information network construction and propulsion

Guiding ideology: guided by scientific outlook on development, directed by intelligent application, based on intelligent industry development, propelled by market demand and innovation; closely surrounding the application requirements in fields like government, industrial development and people's livelihood, strengthening the cooperation with big enterprises like telecom operators and Guangxin operators, quickening network infrastructure construction and providing safe, reliable and advancing basic guarantees for intelligent city development, promoting effortlessly the integration of informatization with industrialization, urbanization, marketization and internationalization; propelling innovation of technology, applied business mode, standards and systems of industry application, improving the abilities of jointly working, cooperatively innovating and market developing in intelligent city information network construction, speeding up the process of intelligent city construction.

Strategic positioning: the construction of intelligent city information network is an important part of intelligent city construction, which concerns core benefits of a country, has a great influence on national information security and conceives huge industrial opportunities. Constructing intelligent city information network is an important way to realize an independent and controllable development of network, and an important content and chance to realize a powerful country through network. Network security concerns national security, economic development and social stability. We should manage comprehensively from national strategy perspective, make proper regulation design, implement policies comprehensively, give full play to the institutional advantage of achieving great with integrated power, identify the difference, increase input, spare no efforts to catch up and make China a network power with clear strategies, advanced technologies, leading industries and both attack and defense abilities.

Strategic targets: making full use of new generation information network technology like internet of things, cloud computing and broadband network to construct intelligent city information network and build city information network infrastructures with the characteristics of ubiquitous intelligence, opening and sharing, heterogeneous integration, seamless mobility, security and credibility, green and energy saving, simpleness and transparency, flexibility and extensiveness, establishing a new generation cloud computing network platform with resource virtualization, computing servitization and management intelligentization; pushing forward the integrated development of 2G, 3G, 4G and WLAN network, building basic information communication network with core characteristics of broadbandization, flattening and integration, propelling two-way transformation of joint developing and sharing of telecommunication network infrastructure and broadcast television network, strengthening the construction of family information network, quickening the scientific research and industrial layout of future network technology. Constructing comprehensive information platform which supports intelligent city diversified information, has the ability of high-performance computing, storing and transmitting, building diversified ubiquitous knowledge and information pivot that covers government administration, public living and industrial development, establishing the operating and administrating "headquarters" and "staff" of the whole city.

Globally, the approach of promoting information network technology development by intelligent city has gained high attention from developed countries. For example, America promoted "Smarter Planet" as national strategy and put forward "national broadband plan". EU issued information society development plan "European Digital Plan" toward 2020, and brought up the plan of speeding up the construction of fast and superfast internet access. Japan released "I-Japan Strategy 2015". South Korea announced "IT South Korea" future strategy, establishing an ICT power in the future integrated field of broadcast communication through constructing the world's most advanced internet of things infrastructure.

2. The developing status and existing issues of intelligent city information network development in China

Generally speaking, although the intelligent city information network in China has made remarkable progress in aspects like technology research and development, standard development and application demonstration, it is still in the initial stage and exists the problems like relatively weak core technology and industrial base, low ability in system integration service and application level, scarce large-scale application and the urgent elimination of industry barriers. Furthermore, the construction of intelligent city information network in China lacks the consideration of information network development mode, positioning, operating and constructing modes. Through the reference of foreign intelligent city information network construction ideology, from a research perspective, we analyze city information network construction status in China and summarize the confronting problems in city information network, which are as follows:

(1) Government administration

1. Information security lacking top-level design: during intelligent city information network construction, information security is very important, but China still lacks the top-level design for information security, weak core technology of information security, low industrialization level, inability in network general defense.
2. Low utilization rate in information network infrastructure: intelligent city construction needs to do specific analysis on existed information network. With the guiding principle of sharing infrastructure, we should quicken the construction of government information resource directory and exchange system, explore and build a multi-level, cross-region, cross-institute and cross-field developing and sharing mechanism of information resource.

3. Weak ability in the deep integration between information network and intelligent city traditional industry: information network's support in information consumption needs to be improved, and the utilization "breadth" and "depth" of information network need to be strengthened. Taking broadband network as an example, during America's construction of broadband network, they focused not only on network coverage and broadband, but also on the integration of broadband network and the real economy like energy and environment, medicare, education, hoping to propel further development in real economy through information network. As far as China is concerned, we need to make progress.

(2) Internet construction

1. Public infrastructure of information network lacks integrated planning, the repetition in construction is severe: the repeated construction in three national operators base station site, pipe network, mobile station tower, premises network is very severe. They lack sharing mechanism and life-cycle monitoring of infrastructure, leading to accident pre-warning and difficult positioning.
2. Weak expansibility of information network system framework: uncontrollable and unmanageable network, lacking guarantee of QoS, low network transmission ability, severe transmission repetition, lacking the function of network traffic engineering, operation maintenance and resource management.

(3) User experience

1. Shortage in network broadband: up till 2012, the rate of actual users of FTTH is less than 20%, with slow development in broadband network and network capacity unable to support a future large-scale application in intelligent city; the latest internet report released by world's biggest CDN provider–Akamai in the U.S shows that China's broadband ranks 75th globally (Q3 in 2013).
2. Severe delay: cross-ISP network delay severely, 80% of the network is in unhealthy condition. In 2012, the global IP total traffic reached 44 exabyte (1018) monthly. It's estimated that in 2016, the global traffic will reach 110 exabyte monthly. The surging traffic makes severe network delay under present broadband condition.
3. Demand analysis of intelligent city information network construction

Intelligent city involves many fields in human's life, including transportation, power grid, medicare, industry, agriculture, tourism, logistics, environmental protection, architecture, and etc. (see Fig. A.4). Each business area has both common demand for information network and specific demand for specific industry.

Considering the supportive effect of information network on intelligent city construction, the construction needs to satisfy the common demand of city

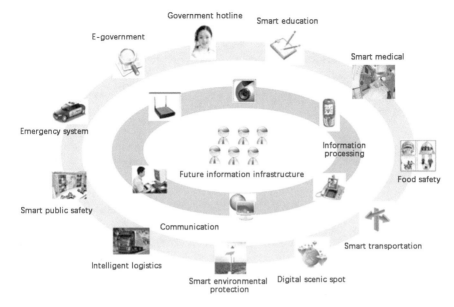

Fig. A.4 Application demand of intelligent city information network

intelligentization, and meanwhile, it has to deeply integrate the specific demand of each industry, so as to realize the deep integration with industries and effective support.

Furthermore, with the integration between information network and intelligent city industries, transforming and promoting traditional industries with new technology, we can change subversively enterprise operating management mode, profit model and subversively change people's working and living styles.

Intelligent City Information Network Total Planning and Construction Contents

The construction of intelligent city information network should closely surround the application demands in fields like government, industrial development, people's livelihood, and build horizontal city information network infrastructures with the characteristics of ubiquitous intelligence, opening and sharing, heterogeneous integration, seamless mobility, security and credibility, green and energy saving, simpleness and transparency, flexibility and extensiveness, and etc. (see Fig. A.5).

Specifically speaking, the bottom-up system framework of intelligent city information network is: sensing layer, network layer, platform layer, application layer, and complete standard system and security system.

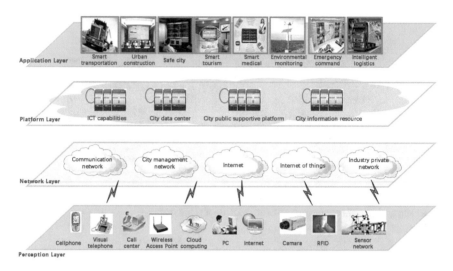

Fig. A.5 System framework of intelligent city information network

The sensing layer is the general condition to realize "intelligence" in intelligent city. The sensing layer has strong environmental sensing ability and intelligence. It can monitor and control infrastructure, environment, buildings and security in a city scale through internet of things like RFID and sensor network, providing ubiquitous and omnipotent information service and application for individual and society.

Network layer is the information expressway of intelligent city, and an important infrastructure of the future intelligent city. The future city communication network should be comprised by optical network with high-capacity, high broadband and high reliability and wireless broadband network that covers the whole city, laying good foundation for city intelligentization. Meanwhile, citizens can surf the broadband internet "anytime, anywhere, as they need", and enjoy broadband services like web TV, HDTV and HD video call.

The core goal of platform layer is to make city more "intelligent". In the future intelligent city, data are very important strategic resource, so the constructing of intelligent city platform layer is an very key part in intelligent city construction. The main goal of platform layer is to solve problems as data fragmentation and unable to share through technologies like data correlation, data mining and data vitalization. Information layer includes data center in every industry, institution and enterprise and city-level dynamic data centers and data warehouses built for data sharing and data vitalization.

Application layer is mainly the various application system built on the basis of sensing layer, communication layer and data layer. The intelligent application layer comprised by intelligent industries, intelligent management and intelligent livelihood promotes to realize the intelligent city of "industrial development, function improvement, better-off livelihood".

The Vision of Intelligent City Information Network Development

The vision of intelligent city information network development is: constructing intelligent city information network infrastructure with ubiquitous integration, open expandability, security and controllability, high-speed broadband, environmental protection and energy saving, providing strong and powerful support for intelligent city service carrying. To build information axis center that covers city management, industrial development and people's livelihood, we can provide comprehensive information sensing, analyzing and processing ability for city operation and management.

Intelligent city information network makes full use of new generation information technologies like internet of things, cloud computing and broadband network to build high-speed, broadband, integrated and wireless new generation intelligent information infrastructure, realizing the interconnection among any people and any things anytime, anywhere. Under this network system framework, whether the user is in front of a computer, in the kitchen, shopping in a convenient store, or waiting at train station, they can get access to the internet conveniently. During the construction of city information, the ubiquitous sensor network will be the most fundamental infrastructure in "intelligent city". It can realize full connection of things and perception in the city through information collecting, monitoring and analyzing of the whole city's systems. In the future intelligent city, information infrastructure will be closely related to and integrated with facilities like water, power, gas and highway through sensor network. They will jointly construct city infrastructure, satisfying comprehensively the connection demand between people and things and finally realizing comprehensively sensing and intelligent decision-making on city operation. They will also integrate and optimize various city resources, improve city operation management and service level and promote citizen's living and ecological environment through the wide connection among every information system, information sharing and cooperative operation.

Intelligent government administration: the government institution uses modern information and communication technology to integrate management and service through network technology; realizes the optimization and reorganization of government organization structure and working process on the level of internet and wireless network; provides comprehensively fine-quality, normative, transparent and international-standard management and service for the whole society, surpassing the restraint of time, space and department isolation.

Intelligent transportation: people can look up information of flights, trains, buses, passenger transport and buy E-tickets anytime anywhere without the limit of time and space; the ticket fare can be paid through cellphone fee charging or cellphone payment account, after the payment, people will receive a cellphone message with a QR code and they can validate their tickets at the ticket exit with this cellphone message and then get on the train or bus or any other transportation; people can look up the situation of road traffic on their cellphone and computer so as to plan better trip route and avoid traffic congestion.

Intelligent public service: people can look up bills of water, gas, electricity, provident fund, social insurance, medical insurance and individual income tax through internet, and pay these bills on the internet; and apply for passports on the internet. Through intelligent city platform, people can get services like consulting, housekeeping, daily supplies delivery, community information, reservation and etc.; can get information consulting services relating to maternal care, antenatal training, infant growth, disease, nutrition, nursing, safety and etc.

Intelligent life: through intelligent city platform, people can get gourmet guides, gourmet review, restaurant pictures, customers review, transportation guides, and get information like restaurant environment and service; through intelligent city platform, people can get services like real-time seat choosing and ticket buying, and the latest movie express; combining with local bonus points resource and gathering high-quality merchants, intelligent city platform provides discounts and holiday sale promotion for customers.

Measures and Suggestions

The present economy in China is facing the opportunity and challenge of transformation. It is important to promote the construction of a network power and construct new generation information network with the help of intelligent city. It is suggested that China quickens its research and construction of intelligent city information network so as to improve our influence in the international market of intelligent city construction, and lay solid foundation for China in the international standardized and normalized construction of intelligent city. Specific suggestions are:

1. **Giving attention to the joint construction and sharing of information network infrastructure**

Infrastructures like central apparatus room, access network room, mobile communication base station and pipes should be included into unified supporting planning. Government should conducts and coordinates operators to integrate various demands and resources, construct and share jointly, unify construction and save resources. Besides, we should also give attention to the life-cycle monitoring of information network infrastructure so as to make sure its security and controllable operation.

2. **Strengthening deep application of information network to the real economy of intelligent city**

Deepening the application of information network to real economies like intelligent city education, medicare and transportation, propelling the reform of information network service technology and mechanism to provide more convenient service and more economic fares for applications in related areas.

3. Giving high attention to intelligent city network security and information security

Now, the global root name server is controlled by countries like America, EU and Japan, which is a great threat to China's information security. From the point of guaranteeing national safety, we should give urgent attention to the network information security and management system in the intelligent city construction of China. For the attention in information security and top-level planning, China should release national overall strategy of network information, and construct safe, autonomous, controllable information network through the comprehensive use of policies and regulations, technology research and development, industrial development and etc.

4. Strengthening talent cultivation

Constructing talent team with high quality, supporting the construction of enterprise talent team. Completing quickly the invention and creation incentive mechanism in colleges and universities and scientific research institutions. Increasing the investment and preferential policy to attract oversea talents to innovate and start businesses in China; quickening the attraction of oversea high-level talents through "The Recruitment Program of Global Experts" and the base construction of oversea high level innovating and career-creating talents. Enterprises are encouraged to adopt preferential policy in the mid-west areas with weak technical force in the matter of talent cultivation and employment mechanism. Completing technological innovation incentive mechanism and improving professional's initiative in autonomous innovation and participation in industrialization of research findings. Constructing and completing talent cultivation mode of industry-university-research cooperation on the base of major special projects and priority projects. Strengthening speciality construction of technology industry related disciplines of new generation information network in universities and colleges and secondary vocational school; reforming the cultivation mode of innovative talents; building a new mechanism of talent cultivation with enterprises participation; promoting the cultivation of innovative, applied and inter-disciplinary talents.

5. Reinforcing nursery finance

Based on the integration of existed policy resource and full use of available fund channel, we can build steady growth mechanism of financial investment, establish special fund for technology development of intelligent city information network, exert our efforts to support major key technology research and development, innovation development project of major industries, industrialization of major innovation findings, the construction of major application demonstration projects and innovation ability, and etc. With the consideration of tax reform direction and characteristics of tax categories, and the characteristics of technology industry of intelligent city information network, we should fasten the completion and

implementation of encouraging innovation, conduct the taxation policy of investment and consumption. Strengthening the combination of financial and fiscal
policies, encouraging financial institutions to increase credit aid to the technology
industry of intelligent city information network. Giving play to the guiding function
of government venture capital investment for emerging industries, expanding the
capital scale, assisting and developing venture capital firms relating to technologies
in new generation information network, conducting private enterprises and private
capital to invest the technology industries of intelligent city information network,
leading social funds to invest the innovative enterprises that are in mid-early stage
of entrepreneurship.

Abstract of 'Intelligent Architecture and Home A Development Strategic Research in China'

The concept of intelligent building has developed for almost thirty years since it
was put forward in 1980s. In China, it has formed certain industry scale with many
high-end office buildings and hotels installing intellectualized systems. But from the
aspects of the covering proportion of intellectualized system installing and the per
capita investment of intellectualized system, there is still a large lag in popularizing
rate of intelligent building between China and other developed countries. There is
much development space for the construction of intelligent buildings. With the fast
development of information technology and electronic technique, IBM brought up
the concept of " Smarter Planet" in 2008. As one part of the Smarter Planet, the
concept of "smart city" has gained attention around the world. As an important part
of urbanization, smart city has attracted attention from many cities. Up to
September in 2013, there are 311 cities in China are under the construction of smart
city or in preparation for the construction of smart city.

As the fundamental node of smart city and intelligent forerunner, the development of intelligent building is the strategic policy we are confronting and need to
confirm urgently under the background of smart city construction wave. Should we
strive to develop it under the cover of smart city construction wave, or discreetly
construct intelligent buildings with the guidance of maximizing investment benefits? The strategic policy will decide whether or not the intelligent building market
and intelligent technology will develop in a healthy and efficient way. Therefore,
there is urgent need in researching the operation status of intelligent building
industry and intellectualized system, and combing development strategy of intelligent building industry according to the present industry status and operation effect
of intelligent building and the demand of intelligent city construction, so as to meet
the demand in new-type urbanization development and energy saving and emission
reduction, and construct intelligent building technology system with our own

intellectual property right, making intelligent building the constituent part of "Made in China" strategy.

Thus, the writer of this report did research on domestic and foreign literature relating to intelligent building, interviewed intelligent building designers, builders, operation and management staff about their understanding and demand in intelligent building, and did field research on the operating status of these building intelligent systems by going deeply into typical office buildings, exhibition buildings, hotels, commercial complexes, school buildings, residences and communities. By this research, we concluded the status in China's intelligent building market, construction and operating management, analyzed the problems and reasons existing present intelligent building industry, brought up solutions and made strategic guideline for the future development of China's intelligent building.

The Status, Problems and Reasons of Intelligent Building

(1) Remote monitoring and remote control functions are what actually take effect in application of building intelligent system. The function of remote monitoring and control can improve work efficiency greatly and reduce demand in labor. For instance, there are hundreds of air hundling units in a building. If workers need to turn on and turn off these devices in local control cabinets everyday, then it needs tens of workers to finish this work for nearly one hour. The function of remote control of the building intelligent system can finish this work in a couple of minutes by one or two workers, great improving work efficiency and lowering the labor cost. This is the most practical function of building intelligent system, and it doesn't cost much to realize this function. So we should promote this function in intelligent system.

(2) Due to the requirement of related policies and regulations and strict examination management, the intelligent security and fir extinguishing systems have played a due role and made satisfying practical operating effect.

(3) Many building owners consider intelligent system as face-saving project for buildings, so instead of caring about the operating effect and maintenance of intelligent system, they care more about the intelligent system installing. As a result, the intelligent system doesn't achieve the due function of improving management and optimizing energy-saving. For example, during the research, we found that many buildings install power monitors, but when checking the power consumption, there is no data. So another energy measurement system has to be installed, which causes great investment waste. There are much more examples in the cases like wrong installation of control valve and unqualified control system. However, the automatic control in power station and industrial manufacturing has high-level localization and good operating effect. In sharp contrast to this, in the intelligent building which demands lower automatic control precision than industrial control, 90% of the building intelligent products adopts foreign brands, and 90% of the intelligent buildings doesn't

achieve the scheduled intelligent function. These phenomena show that the building owners don't care the operating effect of intelligent system. They only concern whether or not there has the installation of intelligent system with foreign brands.

(4) Different intelligent subsystems are incompatible and can't exchange information. Intelligent integration system can control every subsystem. As the development of informatization and intelligent technology, more and more building equipment become intelligent equipment, such as elevator, water chilling unit, water pump and air hundling unit with built-in controller. To solve the problem of the information interaction among different intelligent devices, professional organizations around the world have brought up many solutions, such as BACnet, Lonworks and etc. But in actual building, driven by economic interest, many intelligent device manufacturers don't open their device data. It costs tens of thousands of yuan to get a communication protocol converter to communicate with the device, which causes great overlapping investment and investment waste. Limited by budget, many buildings haven't purchased communication protocol converters, and thus the building intelligent system cannot control the device, which reduces greatly the effect of intelligent system function.

(5) The great investment in building intelligent system doesn't bring due benefits
The reasons causing unnecessarily high investment and large waste are that building owners don't value the operating effect of intelligent system, and don't invest enough to maintain. Another important reason is unpractical aims in design and conducting product selection and design with unnecessary high precision.

Solutions

The chaotic market in intelligent building needs urgent solution, and the huge waste in investment is a great pity. To solve these problems and conduct a healthy development for intelligent building, we need to make clear what is the function intelligent building pursues, how to define clearly the necessary functions through standard specification and design documents, and how to safeguard the realization of the functions during construction, acceptance check and operation, what the input and output is, and how to evaluate investment benefits. Thus, this book gives a standardized function description method and a investment benefits analysis method of building intelligent system which act as technological means to solve the problems above.

The standardized function description method is to divide the function of building intelligent system into five different levels: monitoring, safeguarding, remote control, automatic start-stop and automatic regulating. Aiming at each function, we adopt standardized description method to define clearly and accurately, and quantitatively describe functions like measurement accuracy, system response time, data recording interval and etc. As the common language through stages like design, construction, debugging, acceptance check, operation and maintenance, it can transmit function information. Standardized function design description method aims at being understood by non-control professionals like proprietors and equipment engineers, and meanwhile, as handover document, being able to provide control system design basis for control professional engineers clearly and quantitatively. Besides, the function description documents provide clear and quantitative basis for project debugging and acceptance check, and provide operating gist for operation management workers. So standardized function description plays a role in connecting different professions and different stages, avoiding disjunction.

(1) Monitoring function: it describes the monitoring objects like thermal and humid environment parameter, electromechanical equipment state and automated manual operating mode, and subdivisional energy and resource consumption that satisfy management requirements; in the meanwhile, it describe clearly and quantitatively the measuring position, data sampling method, data information, display position and allowable time delay in each monitoring point. Monitoring function standardized description plays a guiding role in project stage like follow-up design, construction and acceptance check. For example, data related information guides the selection of sensor and actuator; installation site guides the installation of sensor and actuator, and avoids false feedback due to wrong selection of measurement point; allowable time delay is indirect guiding for communication rate.

(2) Safeguarding function: describes quantitatively the physical location, sampling mode, response voltage threshold, corresponding actions, action sequence, allowed time delay and recording requirement of the monitoring point that has alarm system or demand safeguarding.

(3) Remote control function: describes quantitatively the operating position, allowed time delay and recording length when starting and stopping the monitored devices through human-computer interface.

(4) Automatic start-stop function: describes quantitatively sequence start-stop control or actuator state of related devices when starting and stopping devices and transforming working condition. Automatic start-stop can control device's automation running according to usage meter. Contrasting to "automatic regulating" function, it has lower request for the CPU in hardware and has simple software programming, and is easy to realize.

(5) Automatic regulating function: mainly describes automatic control strategy of electromechanical devices or environment parameter, which has automatic control point of information and automatic control algorithm. Point of information describes and defines the input and output in strategy and preinstall point of information, including the sampling mode, data accuracy and etc. Automatic algorithm mainly describes control strategy agent, including the

control strategy name, program triggering acquisition, different actions in different conditions, the expected goal and etc.

The assessment method of investment benefits in intelligent building calculates the specific profit value with net present value method through the analysis on the tangible and intangible investment and return during the initial cost and operation maintenance of building intelligent system. Specific income and expenditure include:

(1) Initial cost (expenditure): including the fees generated in intelligent system design, installation and etc. According to related literature, the investment in the intelligent system of public buildings is around 100–300 yuan/m^2, while the investment in that of residential quarters is around 50 yuan/m^2. With these data and the size of construction area, we can assess the initial cost of building intelligent system.

(2) System maintenance cost (expenditure): according to the cost in subsystem maintenance and service each year and corresponding subsystem point, we can assess the maintenance charge per year of intelligent system. The maintenance charge standard of each main subsystem can be selected according to related references.

(3) Reducing the number of administrative staff (income): the installation of intelligent system can reduce the management of heavy current installation like in-building air-condition, lightening, elevator, and reduce staff like security guard. We can assess the labor cost income brought by intelligent system according to annual salary of administrative staff.

(4) Lowering or increasing building energy consumption (income or expenditure): the installation of intelligent system and strategic optimization of its devices can save energy. We can assess the income brought by building intelligent system according to the result brought by a certain public building in Hong Kong. After the adaption of air-condition system control strategy, they managed to save 6% of the total building power consumption. But the present building intelligent systems in a large scale doesn't bring the energy-saving effect as we expected before. On the contrary, due to unproper control strategy, breakdown in devices like sensor and actuator, increasing cases have seen a higher energy consumption. For example, according to our research, the malfunction of humidity sensor in a certain intelligent building has caused rather heavy fresh air cooling load in air-condition system, leading to 20% more energy consumption in the system. Hypothetically, there are 20% humidity sensors in the system having the same malfunction, then the building air-condition energy consumption will increase by 4%. However, the intelligent system is rather complicated in actual operation, and it's hard to confirm specific energy saving or consumption, so the percentage assessed in the energy consumption income or energy consumption loss in intelligent system is for reference only. More specific data can only be confirmed after researching on more samples.

(5) Invisible income: through level analytic and comparison in pairs, we conduct quantitative analysis on the indicators of intangible income brought y intelligent system, and then get the estimated value after quantizing intangible income. Referring to the nine kinds of intangible income brought by project investment, including innovation ability, product quality, clients, management, union, technology, brand, employees and enterprise environment, considering the practical situation of building intelligent system, we can select and change based on the indicators above and present the advantages that intelligent system could bring to buildings, which is the seven assessment indicators of intangible income (represented by u1–u7):

A. employees: increasing their work efficiency (u1);
B. brand: improving advertising effect (u2); increasing occupancy rate and house-selling rate (u3);
C. service: customers having to wait for a while before getting the certain service, the intelligent system can reduce the waiting time (u4); improving customer's satisfaction (u5);
D. environment: increasing buildings or enterprises' adaptive capacity in living environment (u6); promoting buildings or enterprises' image (u7).

The seven indicators above basically cover main benefits the intelligent system brings to buildings or enterprises. Making quantitative analysis of them, we can get the assessing result of the specific value of intangible income.

Finally, with life-cycle cost analysis method, considering the intelligent system stages like design, installation, operation and maintanence, we calculate the fares and the net present value(NPV) of benefits of building intelligent system in building life-cycle, which can be taken as the basis of investment decision-making of intelligent system.

Development Path and Strategies

Based on the above investigation and analyze, the intelligent building and intelligent housing shall follow the principle of—top level design, market oriented, focus on practical results and reinforce the management. And aim at constructing a building intelligent system that led by the function to ensure safety, strengthen management, and improve work efficiency and save energy.

With the development of information, electronic and data technologies, an increasing number of construction equipments have become intelligent terminals. We need to set up a data standard to help all the intelligent equipments to function on the same system platform. It means that we should construct a building intelligent system platform just like Internet, so that any software can enjoy a free operation on it and don't need to make a protocol conversion. We shall take this opportunity to build an independent intellectual property rights system of intelligent building technology to break the monopoly of foreign technologies. Therefore, the intelligent building will become part of the "created in China" strategy.

The focus of intelligent home's function shall be on intelligent fire control, security, energy metering rather than on the remote control of electrical appliances, curtains, or the full automation of lighting in the house, etc. To automate things such as turn on and off the lights, open and close the curtains that can be realized with little effort will do no good but only to weaken people's own ability. And the full automation of lighting according to the illuminance value will consume more energy than manual control. We need to have a clear identification of people's position and function in the building, and shall not to over-replace the human activity with automation.

The demand for the development of new urbanization requires the construction of intelligent building to be more healthy and efficient to satisfy the need for the intelligent urban construction, energy conservation and emissions reduction. Thus we need to ensure that the design, construction and operation quality of the building intelligent system in order to build a new intelligent construction technology and management system. In this way, we shall change the fact that at present the intelligent building technology cannot be completely applied to the practical engineering application, and solve the problem that the intelligent building system cannot play its function in full drive.

To sum up, the construction focus of intelligent building and household should include:

1. **Intelligent construction (for public buildings)**:

   (A) To build an intelligent system construction and management mechanism that throughout the whole life cycle, and to build a standard for the intelligent construction product and data standard to realize seamless connection to the intelligent city system.

   (B) To develop and promote the advanced intelligent construction system that with proprietary intellectual property rights, and to establish a standardized platform system in intelligent building without protocol conversion.

2. **Intelligent housing (for residential housing)**:

   (A) To build a standard for the intelligent housing product and data standard to realize the compatibility between products and seamless connection to the intelligent city system.

   (B) To establish a technology guideline for intelligent housing, and to lead the development and application of intelligent housing products to a more pragmatic and efficient direction.

Abstract of 'Medical Treatment and Health Care of Intelligent City
A Development Strategic Research in China'

Health care access is an inverted pyramid. With difficult access to quality medical services and aging population, we feel a calling to reform for people; booming global intelligence technology brings opportunity for health care industrial chain with team participation.

Definition and Characteristics

Urban intelligent health care integrates health care information. Intelligent health care integrates information, requires partnerships and prevention. Information integration is built upon digital health care. Hospital departments integrate patients and doctor-based information. Information is integrated from hospitals, communities and families, and from medical institutes, social security and financial institutes. Bio-sensing, genome sequencing and human imaging will be fully integrated; collaboration highlights complete health care system and industrial chain. Parties are involved in responsibilities and interests; health care has a long industrial chain. It involves medical equipment production, circulation, supervision and research, health care education, training, supervision and rating, public health management, maternal and child health care health supervision, disease prevention and control. As internet develops, there's broader and more active participation in disease prevention and control; existing diseases will be treated. Intelligent health care is preventative. With the help of wearing sensor, noninvasive detection, environmental and group monitoring, small trends and signals will be detected and intervened in advance to prevent them from happening. In this regard, we will work to make urban intelligent health care more preventatives, predictable and distinctive with broader participation by establishing an urban service system that is people-first and that integrates medical services.

Comparison in China and Beyond

Countries are distinctive in intelligence development such as health care intelligence. UK and Japan promote small and model towns where people live a peaceful life; Netherlands, with an aging population and less medical service providers, promotes medical facilities with which people live on their own; Singapore feels a calling to promote PHM so that medical information is accessible to the masses who manage health by themselves, as health depends not on hospitals but on people themselves; Canada, a country sparsely populated, pools medical resource together with virtual online medical facilities and clinics, due to lack of funding. Resource is effectively utilized in this regard.

Wireless communications and mobile terminals in China set the stage for intelligent health care development. In China, wireless communications coverage increases to urban and rural population. In terms of mobile terminals, China is now a smart phone-soaked economy. In addition, China has a large absolute demand for intelligent health care: medical treatment needs more intelligence for services; public health prevention needs more intelligence for responses; community health management needs more intelligence for disease prevention and control; health administrative department needs more intelligence for decision making.

Further use of information technology is still essential to health care intelligence. Online medical treatment in the UK, PHM plan in Singapore and medical access in Canada depend heavily on medical system that is unified, effective and available at any time. As information technology develops, medical system reform deepens, and health demand increases, we can tell that this is a critical transition and innovation period for i-city building with intelligent health care. To better respond to health care reform and public health demand, to promote a better health care system, and innovates health care services, we will seek health care intelligence in an integrated and shared system rather than in separated and closed systems.

Health care intelligence in China is distinctive. Compared to other countries, this is the moment for research and use of i-city health care in China. When it comes to urban development, information technology use must be taken into full account, urban layout, environmental protection, food supply and prevention will be studied. It is essential to health and longevity of urban residents and implementation of prevention-first strategy.

Urban intelligent health care building must be innovated in a holistic manner. It is closely connected with health care building. Medicare system, hospital construction and community health care center are in line with macro policies in China. In this regard, health care building in i-cities must be regulated in terms of national strategies. For distinctive intelligent health care suggestions for a city, we must rethink in real time innovation and provide a localized solution.

Urban health care intelligence requires a strong leadership and collaboration. Reform of health care is a long-term and multidisciplinary effort of ministries. In this effort, reform of modern technology and shared interests of relevant institutes and groups must be fully understood. Collaboration contributes to effective reform.

With a sound basis, bottlenecks below still have been hampering the building of health care intelligence in China.

(1) Lack of standards and regulations: Standards and regulations in China in this area have not yet integrated into a complete system. There are few detailed and standard policies on telemedicine, wireless medical security, accountability and insurance reimbursement. The lack of standards and regulations makes telemedicine market difficult to be managed in a standard manner, and makes new

technology difficult to be applied in medical industry. There is lack of legal effect of electronic health records and medical evidence filing regulations. Electronic health record is not filed that goes against a sound industrial development.

(2) High cost of sensors: m-health and tele medicine are sensor supported. However, as most sensors, especially high-end sensors, are imported, cheap and reliable sensors need to be developed in China and used in health care sector.

(3) Little access to electronic medical record: As hospitals compete with each other, and internal and external network are separated, an open digital medical system has not been established for most of hospitals, and electronic medical records are not accessible to all medical institutes. If a patient changes the hospital, he has to do physical checkups again; now electronic medical records are not filed in a standard manner; China still has a long way to go if it integrates and shares electronic medical record.

Intelligent health care building is committed to people's wellbeing and longevity. We work to integrate personal medical information so that each person has active access to his health care information; m-health helps to get medical resource at any time. All these works to deliver a safe and efficient health care. With information technology, policy-driven health care will shift from medical treatment in large hospital to integrated primary health care. Health care is used not to profit but to control cost. Health care is not targeted at individuals but at specific groups. Residents will be more aware of health care by providing relevant education. Individuals will be less dependent on medical care providers but be actively involved in self-health care in a move to make them more capable to take good care of themselves.

Strategic Objectives

We must work to establish a health care data exchange, collection and application and sharing system nationwide that is in line with international systems; we shall build a healthcare cloud service platform nationwide, and support medical care application in cities at all levels; an unified online medical payment system will be available nationwide. By 2020, in a city with a population of more than 500,000, an open health care data exchange, collection and application and sharing system will be available. The system helps to update individual medical care records that are accessible to individuals via equipment and at any time or in any case; we work to establish an urban health care resources interconnection system. In the system, qualified doctors and nurses will be positioned online, patients interact with doctors and nurses and get timely medical treatment feedback; we will look to establish an online payment system that integrates all kinds of medicare insurances. In the system, daily medical care payment can be used for medical care institutes at all levels; we have established a complete medical logistics distribution system. In this system, medicines will be delivered to urban residents in need within several hours; we have established an unified disease prevention and control system that is built on scientific data decision making, emergency response mechanism, web-based first aid system and combined care system for chronic diseases. By 2030, we will establish patient-based medical information transmission and integration system among second-class hospitals nationwide in line with international standards. We will build a unified health care cloud service platform nationwide. With the help of the health care cloud service platform, we will develop secondary data application based on health education, government decision-making and scientific research.

Strategic Task

1. Establish a standard urban electronic health record

Electronic health record is essential to urban health care information construction. In this regard, we must emphasize and push ahead with standard electronic health record building. We will establish a complete and standard electronic health record database that can be accessible to hospitals, communities, clinics, public health institutions that are patient-centered. Patients can also take good care of themselves with the help of the database. In this way can we file and share individual health care information at city level.

2. Develop and adopt practical and advanced health care standards

Health care practice varies from areas. Local health care information systems and equipment developers have a specific geographical advantage. In cities, different software suppliers compete with each other for the same product. However, for medical institutes, patients and local authorities that use this software, incompatible data and application fails to integrate drawbacks of medical care data and systems with sources at different stages. In this regard, to restructure the industry, fully integrate leading industrial resources, develop a sound industrial chain, re-use and interconnect industries, we need to develop a health care information system-based information technology standard, and organize standard evaluation, so that systems will be interconnected in a right and effective manner, and data can be integrated in a specific manner.

In addition, as health care in China, itself, goes global, foreign patient-based medical care is available at home and domestic patients will see a doctor in a foreign country. It is a growing trend that medical care information will be integrated among countries and medical care records will be accessible at any time. On this very note, it is necessary to establish and adopt a health care information standard that is in line with international standards. It helps to fast share research and practice results of developed countries for decades in this domain, and contributes to distribution of medical care information worldwide.

3. **Build cloud i-city and big data center with intelligent service care**

Health care itself concerns many fields, is part of i-city-based operation. It is closely related to population, social security, finance, education, and even transportation. Cloud i-city and big data center are information hub and decision-making center for a city. Health care is an important part. As we work to improve intelligent health care based on the first and second tasks aforesaid, we will establish a system that integrates financial payment and social security, and we will integrate health care information with the help of social security and financial payment. It means this information will be filed in the cloud i-city platform that becomes an urban big data center. It contributes to third-party's keeping electronic health records. In this way, medical institutes can change medical care records that cannot be available before, and any individual and institute have access to this information as they are authorized.

4. **Develop m-health and medical sensor**

M-health works to provide convenient, intelligent and personalized medical services in cities. With these terminals, residents have access to health care services at any time at any place; in this regard, self-service, self-learning, identification and diagnosis will be available; customized services will be provided. Services and advice will be provided according to personal health. We will work to develop m-health, such as health monitoring, consultation and education, remote diagnosis, aided diagnosis and treatment, mobile payment and chronic disease management. We will work to provide a brand new medical care services for urban residents. We will develop wearable medical sensors, such as in vitro data acquisition and feature acquisition sensors. We will develop and scale-produce reliable, easy to use and cheap medical sensors, and build a sustainable business model.

Development and evaluation of health care standard that is in line with international health standards will be on the top of the agenda. We will focus on the development and application of standards such as unique identification, medical information transmission, storage and representation of electronic medical records, medical terminology, classification of diseases, classification and coding of drugs and medical devices. These standards are essential to intelligent medicine, data integration and communication among information systems as well as cloud health care in cities. We will work to deliver cloud service that meet the needs of small medical organizations such as clinic, community hospital and health service station. With the help of a unified cloud service, information will be shared and accessible to medical institutes. It helps to reduce information overheads of small medical institutes.

Policy Suggestion

Localize international health care standard: Establish digital health standards intelligence research centers; develop policies that helps to exchange and share data; lay a solid foundation for developing standards; increase combined efforts of political, academic and technological research and development; develop realistic standards; publicize standards; develop standard training and application system; train interdisciplinary talents in line with medical information standards.

Set up a dedicated office for health care coordination among Ministry of Health, Ministry of Industry and Information Technology of the People's Republic of China, Ministry of Finance and Ministry of Social Security: specific office staff must be fully aware of information technology and health care, and coordinate talents at departments. As government emphasizes and strongly supports it, urban intelligent health care construction will be well underway.

Strengthen information infrastructure: Support enterprises to develop medical sensors and non-invasive testing technology, further develop wireless communications; increase coverage of wireless communications, improve speed of wireless internet. Stable web helps doctors and nurses to provide medical care and patients have access to medical care services at any time and at any place via mobile terminals.

Fund open source health care software: Support R&D and application of open source medical software; establish open source health care software code database. As health care is localized, software developers nationwide will expand their presence into health care software system development with certain geographical relationship. As medical software involves many businesses and less key technologies, there is growing redundant development for software suppliers nationwide. With fierce competition, medical software developers get few profits and have difficulty in system maintenance. To optimize industrial structure, restructure top resources in the domain, we will work to have a sound industrial chain, reuse software industry, and improve standards. With the help of R&D of software system of open source health care, we need to provide low entry costs for start-ups, and make software system more sustainable, and reduce redundant investment.

We must work to accelerate development of supporting laws and regulations. Legal construction of health care information must be improved. Issues concerning laws and regulations in an information age of health care must be properly dealt with. Information security and privacy protection must be well dealt with. Put in place and improve laws and regulations on electronic evidence and evidence collection in health sector. We must work to create a sound legal environment in an information age of health care. We must work to improve mechanism building, and encourage the development of health care information standards. We will ensure continuous development of planning, development, and promotion and operation management of health care information standards. We support issuing, implementing and promoting a unified and standard health information system. We will actively encourage the development of international health information standards.

We have established health care information standards that meet actual demands, in a move to prevent information isolated island in health care sector. We will provide extensive advice on and promote collaboration of health care information and drive relevant industries forward.

Abstract of 'Security of Intelligent City A Development Strategic Research in China'

Definition of City Safety

With a growing need for city safety, city safety becomes more important. City safety is a complex and comprehensive effort that involves a lot of fields. How to ensure a safe and happy life for residents in an effective and sound manner have been a common concern in the academic, engineering, management and political circles.

What is city safety? City safety means how city runs and functions are risk free. As i-cities develop, cities are managed in a more intelligent, collaborative and holistic manner. Definition of city safety also becomes more intelligent, collaborative and holistic.

Given urban development in stages, definition of city safety has always been improved. I-city safety involves city safety in all stages and a series of other issues that arise from intelligence such as mechanical malfunction, information disorder and system crash. As definition of city safety evolves as times goes by, we put people first and highlight safety. Intelligence and safety are people-centered. All other factors are also people-centered. It suggests that, city safety is dedicated to safety of people and things as social entities.

Security threat such as crime, natural disasters, technical disasters and terrorist attacks becomes more salient for modern cities. Crime and natural disasters are traditional security threats, while technical disasters and terrorist attacks are new threats. With urban sprawl and globalization, these security threats have new trends.

Study on Development of City Safety in and Besides China

1. Study on city safety in and besides China

In western countries with series social problems, they focus on theoretical research of personal safety defense and design. In this regard, definition of safe city is more defensive.

"Safe city plan" was initiated from the UK in 1988. It was primarily set to reduce crime, reduce fear of crime, and create a safe and prosperous city with flourishing economy and community life. Afterwards, the campaign became a worldwide effort. In 1980s, the United States developed concepts and theories by public safety circles and used by western countries. In Asia, with special geology, Japan often lies in harm's way from natural disasters. As a result, the country took the lead in studying city safety that focused on disaster reduction.

From late 1990s to early 2000, people had a strong interest in city safety. In 1996, for the UN's tenth anniversary of international disaster reduction efforts, the UN determined international day for disaster reduction with a theme of urbanization and disaster. On October 5, 1998, world habitat day, the UN developed a theme of a safer city. In 2001, after the terrorist attack in the United States, response to terrorist attack was added into the list of city safety study that used to focus on disaster, war air raid and safety crime reduction. Countries came a long way in improving city safety; ensure life safety and social stability.

2. Development of city safety in and besides China

(1) New York-application of database and communication system

With the growing use of electronic computers and internet, with rapid development of information technology, grid management, internet of things, cloud computing, information sharing and communication has heated up. These high-techs underpin safety of New York. Early in 1994, New York City Police Department created a new police model, known as "Compstat". This is a police model created by New York City Police Department to sort out crime data with computer technology timely, to develop electronic crime maps, to analyze crime patterns and trends, to provide advice on optimized police resource allocation and clear police account-ability. In May 2010, New York City Police Department worked with IBM, as IBM was capable of collecting; sharing and dealing with information, the police department effectively makes the most of data to push ahead with case solving. IBM and its partner Cognos co-established a real-time crime database, so that New York City Police Department was more actively an effectively involved in cracking down on crimes.

"Network command" system, an effort of New York Fire Department, is another example of mature application of database and communication technology. The network command composed of sound, picture and data built on information sharing got beyond conventional wisdom, and was seen as a new way of emergency command.

(2) Stockholm-Complete urban industrial chain

Sweden's largest transportation company, SL, invested £ 25m in building a public transport safety system. More than 15,000 web cameras were installed. Cameras, alarms and warning systems were connected to the central security station. City safety system could not do without strong support from information and commu-nication technology (ICT). ICT clustering was developed at Kista Scientific Park in eastern Stockholm. Since 1990s, Kista has been a global information communi-cation industrial clustering with wireless communications directs the effort. Many world-famous multinationals such as Ericsson, Nokia, IBM, SUN, Oracle, Intel, Compaq, Motorola and Microsoft established R&D centers or production bases at tech city.

(3) Beijing

Beijing focused on industrial development of internet of things and application of city safety and emergency management. Internet of things was applied to emergency command supporting. Real-time sensing was promoted; information sharing and intelligent analysis were promoted. Dynamic monitoring of city safety, intelligent research, and emergency field sensing and rapid response were improved.

3. Enlightenment

(1) We must work to establish a management system for city safety. We must strengthen innovation of relevant systems, put in place and improve city safety management laws, so that the urban safety will be managed in a legal system.

(2) Great effort must be made in the construction of public information network. We must work to establish unified disaster database and emergency information management system.

(3) We must strive to establish a local safety management system and traction through large systems. We must promote city safety in large systems and projects.

Analysis on Development Demand for City Safety in China

1. City safety characteristics in China

Firstly, a city has a huge and concentrated population but only a small amount of per-capita resources. Secondly, a city is chain structured, with single and weak chain. Thirdly, urban system is sensitive to natural or man-made interference and change. There is lack of dynamic restructuring and self-recovery. Fourthly, there is imbalance between social capital and social relations as China grows stronger.

2. Analysis on development need for city safety in China

City safety building consists of 3 parts, that is, improving management systems, building innovation systems and organizing industrial systems. Now China has a well-developed management system, but poor safety innovation and industrial systems. Reasons behind it are: firstly, lack of an innovation system with state-level technical innovation platform as the core. China exerted itself to promote technology that underpins national security and public safety, however, breakthrough of the technology requires technical innovation system that is yet to come into being. Secondly, lack of industrial system with industrial clustering as the feature. Now, there are a great number of small enterprises engaged in city safety. There is lack of leading enterprises in this domain. There is lack of innovation. Imitation, copy-catting and OEM prevail in the market.

Overall Strategies of Developing and Promoting City Safety in China

1. Overall objectives

In the framework of the overall national security, we should actively respond to urban traditional and non-traditional safety, physical and virtual security. We should be information driven and service-oriented. We should build a city safety capacity system for scientific early warning and effective prevention and control, in a move to deliver happy life in a better city.

Objectives consist of 6 parts:

(1) "Safe grid" with strong and comprehensive prevention and control;
(2) "Organic structure" with disaster response and recovery capacity;
(3) "Emergency system" with peak capacity;
(4) "Eco layout" emphasizing environmental protection;
(5) Healthy city for high-quality life;
(6) "Information security" with intelligent prevention capacity

2. Overall strategies

In response to improve management system and management capacity, we should actively respond to urban traditional and non-traditional safety, local and overall safety, physical and virtual security. We should be information driven and service-oriented. We shall open a pathway to integrated technology, service, capacity and systems.

In response to improve management, we shall establish city safety building systems in which technology, industry and management will be improved in an overall manner; we shall establish city safety management systems, so as to integrate evaluation, planning, development and operation; we shall establish an innovation system that is efficient and flexible, and in which talents, platform and systems are well interacted. We shall establish a service system that consists of residents, communities and cities.

In terms of management innovation, we must be information-oriented. We shall share information, remain information-driven and service-oriented. We shall shift from system integration to capacity integration. We shall shift from "made in" to "made for". We shall shift from providing data to providing information. We shall shift from governmental building to service purchase.

3. Development task

(1) Put city safety on the top of the agenda of development

To remain committed to people-first outlook on city safety development, firstly, we need to be aware of safety need from the masses, and always put people's rights on the top of the agenda. Secondly, we need to be aware of growing destruction of people as cities develop. We need to have safety awareness that is forward-looking. We need to implement a high standard protection for city safety.

(2) Build a city safety management system

We need to develop a city safety management system that consists of evaluation, planning, development and operation.

For city safety evaluation, we need to focus on safety evaluation in key areas, key infrastructure evaluation and safety evaluation of important activities. We need to set up a comprehensive assessment institute for state-level safe operation, and establish city safety operation and supervision indexes. Priority shall be given to monitoring and control planning, early warning and forecast planning, emergency decision planning, emergency response planning and emergency management planning. For capacity building, we need to improve flexible management and peak ability of emergency response. In terms of city safety operation, we work to build comprehensive city safety system operators.

(3) Build a forward-looking city safety technical innovation system

To build a technical innovation system for city safety, firstly, we need to establish an early warning system of comprehensive monitoring. With an integrated and interconnected sensor system in multiple directions, we work to get continuous and complete data; secondly, we need to build the peak ability of disaster response, that is, national public security emergency network. The network integrates multiple businesses and modes. The unified, specific and stable network responds rapidly nationwide. The system consists of 1 communication network+1 "big data" processing center+1 emergency response network center. Thirdly, we will establish an integrated information emergency publishing platform in a reasonable manner, so that information with sources will be integrated, multi-frequency information will be issued, and governments are integrated.

(4) Build a service system that is information-driven and practical

Firstly, we need to shift from "made-in" to "made-for", that is, we need to depend not on manufacturing but on manufacturing customer-first products. Secondly, we need to depend not on providing data but on providing information. In data acquisition, transmission and storage system in the future, data will not be delivered to end-customers in an easy manner. Data will be effectively exported intelligently in all links of the data. Thirdly, we need to depend not on departmental network but on safe information grid, that is, departmental networks, as roles are fully played, with the help of interaction and integration, need to establish a city safety information grid in which information flows, resources are shared and functions support with each other. Fourthly, we need to depend not on project construction but on operation, that is, we need to expand our reach into maintenance, management and operation as well as construction of the city safety system with internet of things.

Suggestion on City Safety Development

1. Put city safety on the top of the agenda of urban development

We look to establish an assessor for safe operation of state-level urban systems, and establish indexes of city safety operation monitoring. Besides, we must remain committed to people-first city safety. Firstly, we need to be fully aware of safety needs from the general public. We need to always put rights of the masses on the top of the agenda as we seek urban development. Secondly, we need to improve safety awareness that is ahead of our time. We need to exercise high-standard city safety protection system. Relations between safety and development will be coordinated in line with scientific outlook on development.

2. Innovative social management; depend not on accident disposal but on risk prevention and control

By establishing a comprehensive management strategic mode, we ensure efficient coordination, comprehensive integration and management profits. We need to ensure that responsibility is governed by law, risk evaluation is in line with systems, risk is always monitored, management institutes are well integrated, and emergency responses are integrated.

3. Establish a forward-looking city safety technical innovation system

It is suggested to set up a national public safety emergency network at the state level. We need to increase investment and make major breakthroughs, and increase independent innovation involving technical R&D, system integration, standard protocols and construction, which underpins city safety.

4. Promote city safety industrial development

City safety needs to be promoted by industries. Now there is lack of industrial clustering-driven public safety industrial systems. It is suggested to develop national urban safety demonstration area and drive industries forward in areas with comparative advantages.

In addition, city safety needs to be underpinned by operation. It is suggested to promote pilot safe cities, train professional city safety operator and service provider. We need to explore an operation mode for safe cities.

Abstract of 'Environment of Intelligent City A Development Strategic Research in China'

Connotation of the Intelligent City Environment Development

Intelligent city environment development is the use of high-tech information technology for the collection, processing and application of the environmental information. To form a ubiquitous environmental information transmission network by using of radio frequency identification, wireless transmission, networking and other technologies; to realize environmental integrated intelligent management with the goals of environmental quality improvement and environmental risk prevention and control by rapidly screening, processing and evaluation of environmental information through the information center, cloud computing platform, etc... The connotation of intelligent city environment development can be summarized into 3 aspects:

(1) Enhance the ability to access massive complex environmental information. To promote the ability to access and transmission of environmental information through the construction of ubiquitous environmental information to identify the monitoring of the neural network and the Internet of things, which can provide data support for intelligent city environmental management.

(2) Improve the integration ability of multi-source and multi-dimensional environmental information. Integrate the multi-source and multi-dimensional environmental information through the distributed database. Establish the of intelligent city environmental monitoring and information service data exchange and sharing center, to improve the integration capabilities of multi-dimensional, multi-source, multi-scale, massive environmental information.

(3) Build multi-functional and integrated environmental information processing application platform. The evaluation, simulation and operation of environmental information are the core of intelligent environmental decision making. Build a cloud computing platform against the characteristics of complex and massive environmental information, to provide a comprehensive and multi-functional integrated solution for intelligent city environmental management.

Achievement and Main Problems of China's Environmental Information

China's environmental management information construction started late in 1980s. With the national emphasis on environmental protection work, the intensity of the construction of environmental information has been increased, and the construction of information system and infrastructure capability has been widely carried out.

(1) The construction of environmental information capacity develops rapidly. Built a multi-level environmental protection information agency for information center of Ministry of Environmental Protection, 32 provincial information centers and more than 100 city environment information centers, widely carried out environmental information basic software and hardware facilities, environmental protection business database, information systems and other construction works at all levels of business departments and built a two levels of environmental protection private network of national and provincial, all levels of environmental protection system Intranet and the Internet.

(2) Environmental monitoring sensing network has been primarily built. After more than 30 years of work, China environmental monitoring master station and its subordinate stations has primarily built China's environmental monitoring sensing network, realizing the monitoring to environmental factors such as atmosphere, water, ecology and key pollution sources. Built 365 pollution source monitoring centers at two levels of provinces and cities, carried out the automatic on-line monitoring to more than 150,000 key pollution sources by using Internet of things; built a perfect city environmental air automatic monitoring system in 113 cities; built the national surface water environmental quality monitoring network, centralized drinking water source water quality monitoring network for cities at prefecture level and above, national surface water environmental quality automatic monitoring network and etc.

(3) The construction of business system has begun to take shape. All levels of business departments widely carried out the construction of environmental protection business database and information system, a batch of business system construction formed scale, such as environmental quality monitoring, pollution source monitoring, environmental emergency management, sewage charges, pollution complaints, approval of construction project, nuclear and radiation management, which has provided strong support and services for the realization of China's environmental information, scientific management.

However, due to the lack of overall planning and top-level design in the process of environmental information, development has encountered many problems, mainly including:

(1) Lacking overall planning, the construction of standard system lags behind. The construction of China's environmental information is lack of overall planning and guidance of the national level, the business sector acts of their own free, overlapping investment, blindly construction is very common. And some information systems only focus on the initial investment, lacking of continuous operation maintenance, resulting in a waste of the system. On the other hand, the construction of China's environmental information standard system lags behind, and there is a lack of guidance and norms for environmental information work, which unable to cope with the situation of rapid development of information technology. Especially in the face of the challenge of environmental protection big data, lacking of the core technology, standards and management mechanisms of environmental information security and application.

(2) Poor data reliability, low level data share The public has a strong demand for accurate and transparent, timely and open environmental information, however, environmental data lacks of standardized regulatory process in monitoring, collection, processing, and the data quality is affected by many factors, which makes poor data reliability and some queries in data quality. Problem of "One data for multiple sources" caused by multiple departments and multiple exports, such as the problem of environmental statistic data, pollution census data and several sets of data from the pollution monitoring originated from multiple departments and inconsistency between data, so it can easily cause data confusion in the practical application. As a kind of important strategic information resources, the contents of environment data involve the field of resource and environment, society and economy, covering multitudinous departments, such as environmental protection, water conservancy, meteorology, land, forestry, agriculture, due to the monopoly, closed of multi-level business system settings and information resources between the department, forming a situation of "Longitudinal information silo, horizontal information isolated Island", and it is difficult to share data, environmental data resources are not fully exploited and utilized effectively.

(3) Resource dispersion, low application level of informationization. Due to the lack of planning, business system built by various departments is dispersed, universality is poor and linkage is weak between the systems. And currently, most of the business systems mainly achieved the basic functions, such as data management, inquiry, statistics, which lacks of environmental big data simulation application, spatial visualization analysis, data mining, and intelligent decision-making and other environmental information depth service capabilities.

Application of Big Data in City Environmental Management

Big data will play a role in the Internet of things monitoring data, resource sharing services, intelligent environmental management and decision-making services in the field of environmental protection.

1. Big data's role in environment-friendly internet

As environmental monitoring system is improved in China, automatic online monitoring system is widely applied in sectors of pollution source, water environment, atmosphere, acid rain and dust storm monitoring, and satellite remote sensing is widely used in water ecology, air quality, straw burning and regional ecological change monitoring. A lot of environmental monitoring data is generated every day. As environment-friendly internet technology is promoted, with the help of GPS, satellite remote sensing, video surveillance, infrared detection, radio frequency identification and wireless network, we need to establish an all-in-one monitoring network around the clock that can be found anywhere and at any time. Data of kinds of environmental awareness equipment increases fast. Traditional data storage, analysis and handling technology fails to respond to huge and complex data. Big data technology came into being. Big data is based on large control centers and mobile terminals, and built on cloud storage, large-scale distributed computation, cross data mining and analysis. It contributes to online access, on-demand access, real-time handling and model analysis. It helps to accurately and timely get effective environmental quality. It helps to lay a data basis for local environmental authorities to organize overall control, pollutant discharge regulation, environmental law enforcement, and provides enterprises with optimized production and analysis on energy conservation and emission reduction, so that emission will be reduced and environmental risks will be prevented.

2. Application of big data in the integration and sharing of environmental data resources

It is the inevitable direction of China's environmental management with carrying out resource sharing and business collaboration by each department to deal with environmental problems. Environmental information sharing and service based on the big data technology will be the premise and technical support to realize the management of complex, diverse, and massive environmental elements. Big data technology will integrate all kinds of environmental data resources distributed in different monitoring systems, information systems and sharing platforms combining with the cloud computing data center, to build a big data management platform of distributed storage, virtualized centralization of management and scheduling, to provide a unified portal and one-stop search interface and realize the management of cross-region, cross-department and cross-platform to the environment data. To integrate the environmental information resources, and construct different types of environmental information integrated using model on this basis, to provide data for different users and to meet the needs of all kinds of environmental management and information sharing by cross and excavate information in different data sets, different servers and different data nodes.

3. **Application of big data in intelligent environment management and decision-making services**

The final use of big data oriented the application and service, to build an intelligent decision-making and service system of big data in the intelligent environmental protection, which can provide environmental quality monitoring, pollution source monitoring, environmental risk evaluation and early warning, emergency scheduling, supervision and law enforcement and management decisions and other services for environmental management, to realize intelligent management all the way of the environmental protection business. As a resource and a tool, big data relies on Internet of things, distributed storage, cloud computing, data mining and other technologies, excavating information products of high value and diversification from the massive, heterogeneous, multi-dimensional, distributed data, to realize intelligence, fast acquisition, processing, decision and feedback of environmental information, which can make the environmental management with more intelligent decision-making power and insight and promote the improvement of the intelligence level of environmental management.

Strategic Target and Main Task

1. **Strategic target**

With the starting point of solving serious environmental problems in Chinese urbanization, relying on the city's high informationization, to establish ubiquitous real-time environment service system of sensing, integration, processing, decision-making and service for city environmental information. With the construction emphasis on intelligent sensing, intelligent processing, intelligent application of environmental information, to build intelligent management system of city environment, to promote city environmental protection changes from pollution source control to the environmental quality control, from target gross control to capacity total control, from passive emergency management to active risk management, to achieve the goal of harmonious development of city environment, economy and society.

According to the situation of the development of the existing intelligent city and the national economy and social development plan, the strategic objectives of the intelligent city environment development are:

In 2020: key pollution sources monitoring system will be built in all cities at prefecture level and above, environmental quality on-line monitoring system will be built in 80% cities at prefecture level and above and 50% county-level cities, 3–5 remote sensing systems of pollutions in ecology, atmosphere and water body for major city agglomeration will be built, form environmental monitoring network of completed environmental monitoring elements, all-around monitoring method by collecting ground-air-space; environmental information will highly integrate into city environmental management, information system of intelligent environment integrated environment risk early warning and forecasting, emergency feedback, optimization control, decision supporting will be basically completed.

In 2030: key pollution sources and environmental quality on-line monitoring system will be built in all cities at prefecture level and above and 80% cities at prefecture level and above, 10–15 remote sensing systems of pollutions in ecology, atmosphere and water body for major city agglomeration will be built; strengthen the construction of data sharing platform of city level, establish 100 big data integrated processing centers covering environmental protection, water conservancy, land, meteorology, agriculture and other departments, form the international leading intelligent environment application system.

2. **Main tasks**

Build the stereoscopic city environment sensing physical space of all-weather and all-dimensional, construct key pollution sources monitoring information system, informational ground environment quality monitoring station network, high resolution hyper-spectral remote sensing monitoring network and other information intelligent sensing infrastructure. Promote the application of Internet of things, laser communication, global information grid, cloud storage, cloud computing, virtual reality and other high-tech information technology applications, develop the construction of environmental information sharing platform and big data center, guide the formation of cyberspace of the national environmental information management and optimization of environmental decision-making. Give full play to the service function of city environmental information, and construct the information system for integrating city environment early warning and forecasting, emergency feedback, optimization control, decision supporting and other functions. Strengthen the sharing and open of environmental information, improve the openness, objectivity and comprehensiveness of environmental information, and guide the whole society to participate in the mechanism. To improve the supporting ability of environmental information to city environmental management, and to promote the development of environmental management mode, this aims at the environmental quality improvement and environmental risk control, thus supporting the establishment of eco-civilized city.

Key Projects

1. **Information engineering of i-city ecology protection**

The Twelfth Five-Year Plan for Informatization of National Political Affairs lists informatization of ecology protection into 15 state-level information system engineering. The project, led by Ministry of Environmental Protection, is a joint effort of National Development and Reform Commission, State Forestry Administration, Ministry of Agriculture, Ministry of Land and Resources, National Bureau of Statistics, Ministry of Industry and Information Technology of the People's Republic of China, State Oceanic Administration, Ministry of Water Resources, Administration of Quality Supervision, Inspection and Quarantine, State Meteorological Administration and National Energy Administration. It works to generally share information such as pollution sources, pollutants and ecological quality, improve environmental management in key river and regions, and effectively enhance monitoring and assessment of environmental ecology and biodiversity conservation. It works to prevent industrial pollution and promote environment-friendly society building. In response to salient environmental problems that threat health of the masses, to avoid ecological deterioration at the source, we need to make the most of internet of things and remote sensing, and improve ecology protection system that involves soil, forest, wetland, desert, ocean, surface water, groundwater, and atmosphere. With a new generation of information network technology, we need to dynamically bring together industrial enterprise pollution monitoring information and strengthen industrial pollution and greenhouse gas emissions assessment and monitoring.

In the construction and promotion of China's intelligent city, forming a national level ecological environment protection information engineering with ecological environmental protection information engineering as the major project, environmental protection departments and other ministries work together, share and co-construct, to break down the barrier and to realize the interconnection. Most applications of government informatization at all levels are still rest on levels of information release, office systems, convenience centers, etc., so it's very difficult to realize the information sharing and business collaboration across departments. Difficulty in information sharing and business collaboration is seemingly a technical problem, in fact, is a question of management, moreover, is underlying problem of system and mechanism. Integrate information technology within the environmental protection system with the aid of "ecological environment protection information engineering", undoubtedly, this is a very good opportunity. With this chance, change the situation of independent construction, closed construction, self-contained system to build together across departments. Strengthen departments' cooperation, promote united office, collaborative business, promote the fine management of pollution sources with the aid of ecological environment protection information engineering, to gradually form the "integrated" government departments join forces, which can significantly increase departmental administrative efficiency.

2. Intelligent environment sensing physical space and manage cyberspace of national level

The core of the intelligent environmental protection is the environment big data of high quality, the ubiquitous sensing of the environmental information is the basis for realization of environmental protection intelligentization, and the key is to solve the problems existing in China's environmental sensing network infrastructure construction, such as the independent dispersion, blindly repeated, poor universal linkage and so on. First, it requires to strengthen the deployment schedule of overall planning and top-level design, scientific deployment, unified construct intelligent sensing infrastructure, to form pushing pattern of equipment compatibility, information unobstructed, resource sharing. At the same time, the construction of intelligent sensing infrastructure needs to combine the city's own characteristics and development strategy, focus on solving the outstanding environmental problems of the city, to reflect the city's characteristics and avoid "one side thousand cities". Second, to build environmental information real-time sensing system of integrated ground-air-space, and construct key pollution sources monitoring information system, ground environment quality monitoring network, remote sensing monitoring network and other infrastructure. Take full advantage of RFID, QR code, GPS, GIS, sensors and other information technology, sensing, measuring and transmitting environmental information anytime, anywhere.

Intelligent environmental protection needs to focus on the breakthrough of current situation of application of existing environment information low-end service, which must rely on a new generation of computer technology and virtual space technology, build a virtual digital space using environmental protection big data to optimize cyber of the environmental protection strategy, solve the bottleneck problem of city environmental protection. First, collect multi-dimensional, multi-source information resources, link environmental protection, water conservancy, meteorology, land, forestry, agriculture and other departments, to guide the management of information changed from the "lonely island type" to "borderless type". Second, build environment big data processing center, comprehensively apply the cloud storage and cloud computing, distributed database, analogue simulation and other technologies for the integration management and comprehensive treatment to the environmental protection big data. Third, build the environmental information comprehensive decision-making service platform, to achieve environmental quality early warning and forecasting, emergency feedback, optimization control, decision supporting, open sharing and other functions, to achieve the city's environment quality intelligent management.

Abstract of 'Intelligent Commerce and Finance A Development Strategic Research in China'

Business and financial activities based on Internet have substantive characteristics that are different from traditional economic activities.

Original Background and Connotation of Intelligent Business and Finance

1. Original background of intelligent business and Finance

(1) Needs of servicing "Four Modernizations Synchronization". Intelligent business and financial activities is an important support and platform to achieve the "Four Modernizations". With the technical support of informatization, networking, mobile informatization, and in the course of building "Four Modernizations Synchronization", intelligent business and financial construction is the ideal catalyst and construction link, which can vigorously promote the development and deepening of industrialization, urbanization and agricultural modernization while effectively promote information construction.

(2) Needs to greatly improve the level of information in business and finance. Inner demands of the informatization for the business and financial industry become more urgent. In-depth informationization of business and finance is the core of national economy and social informationization. And the industrialization of information and the informationization of the industry are directly related to the development and quality of information.

(3) Meet the needs of the network economy rapid development. The network economy is not a pure "Virtual" economy which is independent of the traditional economy, and is completely opposed to the traditional economy, which is an advanced economic development form, actually produced on the basis of traditional economy and improved by the modern information technology with the computer as the core.

(4) Needs to improve the function of physical city and promote the development of virtual City. Currently, China's urbanization is facing enormous pressure of population, resources, environment and society. With the development of modern information technology, especially the Internet technology, modern network has formed the second space of human gathering, the virtual space is parallel to the physical space of the earth's surface, and the volume also increased sharply. Virtual space will increasingly undertake the function of the city.

(5) Meet the needs of business model reform. Modern business model has the characteristics of knowledge intensive, technology intensive, capital micro-scale type, asset light and thin. Emerging information technology drives business model reform. Currently, social structure and consumer attitudes are changing, the consumption structure has been accelerated, and the younger generation has gradually become a new consumer groups, consumer demands are more diverse, personalized, information channels appeared diversification (including portals, social networks, mobile phones, television, traditional media, etc.), shopping channels diversification (physical stores, TV shopping, mail order, e-commerce, mobile commerce, etc.), delivery modes diversification (shop logistics, small logistics, group

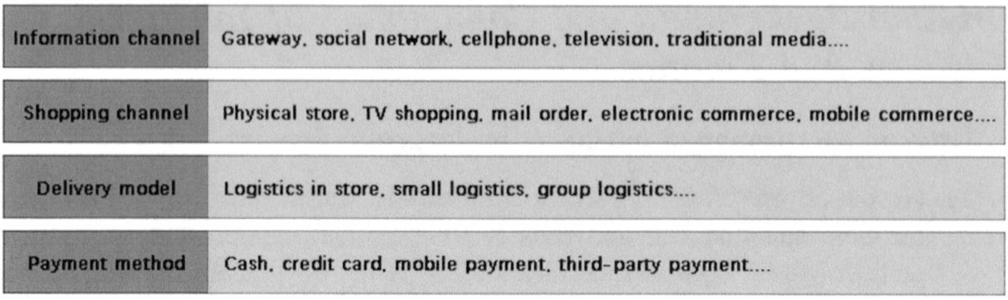

Information channel	Gateway, social network, cellphone, television, traditional media....
Shopping channel	Physical store, TV shopping, mail order, electronic commerce, mobile commerce....
Delivery model	Logistics in store, small logistics, group logistics....
Payment method	Cash, credit card, mobile payment, third-party payment....

Fig. A.6   Diversified services available for the consumers

logistics, etc.), method of payment diversification (cash, credit card, mobile payment, third party payment, etc.) (see Fig. A.6), meanwhile, consumers are not only satisfied with the diversified services, consumers pay more attention to consumer experience and services of the unified channel, hope to provide one-stop service, hope to get fast, convenient and intelligent business and financial services with rich in channels.

2. The Connotation of Intelligent Business and Finance

The existing theoretical research and application is still in its initial stage, therefore there is no clear definition to the connotation, industry scope, operation mode, business model and other key factors of intelligent business and finance.

(1)   The Connotation of Intelligent Business

IBM believes that with the development of new technologies such as Internet, cloud computing, social media and mobile commerce, the business environment has changed significantly. Companies have to create a new business model by employing new business intelligence and other technologies, so that business activities are performed in a much more intelligent way.

Currently, a consensus has not yet reached on the definition of intelligent business, but the various definitions can be divided into two categories: academic and entrepreneurial. We believe that Intelligenter Business is the intelligent transformation of all business activities, that is, a continuous intelligent transformation and improvement process to the decision-making and management of various organization and individual business activities, transaction achievement and fulfillment by comprehensively employing a variety of modern information technologies and management methods. In short, it is an ever-evolving digitization, informatization and intelligentization process for all of the production and service business activities.

Intelligent business involves e-commerce and business running from purchase, inventory, sales, storage, and logistics to settlement, financing and decision making, management control and transaction in financial market. Business intelligence involves   e-commerce   (including   cross-border   e-commerce),   supply   chain management (such as intelligent logistics), business services, industrial regulation, e-government (commerce department). Business intelligence lies in innovation of management, operation and profiting. Intelligent business is an evolution.

(2) The Connotation of Intelligent Finance

From the view of intelligent city construction, we think that "Intelligent Finance" is an important part of intelligent city construction, which is also the advanced development stage of financial informatization. During this development stage, a high degree of integration of modern information technology with organizational structure, business processes, product development, customer service, customer experience, risk management and other areas of the financial industry, is a kind of financial system operation theory that aims to maximize meet customer needs, is a kind of financial system operation mechanism that takes "massive information— knowledge accumulating—scientific decision" as the core, and is a kind of financial system operation pattern whose core goals are to be more stable and effective. Specifically, it can be divided into the following five aspects:

1. To timely respond to customers' financial business needs through dynamic IT infrastructure;
2. To improve the financial transaction decision support capabilities of financial institutions and investors through intelligent analysis and optimization of massive data, high-frequency data and big data;
3. To provide customers with personalized, convenient financial products and services through flexible perception of changes in customer behavior patterns;
4. To develop safe, convenient and fast financial payment system through the interconnection between financial systems and other platforms, self-help counters, mobile platforms, networks, mobile banking and other electronic support;
5. To strengthen the internal risk management of financial institutions by applying various tools to avoid financial risks.

3. Intelligent Business and Financial Construction

(1) Construction of intelligent business. Intelligent business applications and construction architecture are shown in Fig. A.7.
(2) Construction of intelligent finance

The core of intelligent finance is not only timely, completely acquiring information and quickly, efficiently, extensively sharing information, more importantly, realizing the real-time processing and intelligent analysis through technological innovation. Mining the effective information needed from a bulk of banks, securities, insurance and other financial data, and finding the valuable financial knowledge, to support and make the reasonable and efficient financial decision, and achieve financial intelligence. On the basis of the perfect integration with the new technology, the intelligent finance is separated from the strict matching requirements of material flow and capital flow by traditional financial activities. There is

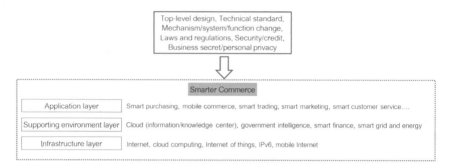

Fig. A.7 Intelligent business applications and construction architecture

big difference in management philosophy and operation mode between intelligent finance and traditional financial activities, as shown in Table A.1.

Overall, there is big difference between intelligent business and finance and traditional business and finance (see Fig. A.8). The main construction body includes both government which serve as a functional department and a public service department, and manufacturing enterprise and service enterprises that provide software, hardware, network communications, banks and other services, as well as the enterprises, families and individuals that serve as the main body of consumption and needs. The information technologies needed to use include data warehouse, data mart technology, data mining technology, OLTP, OLAP, advanced analysis technology, virtualization, data visualization, big data, computer network and WEB technology, cloud computing, internet of things, etc. The Management methods

Table A.1 The difference between traditional finance and intelligent finance

Serial number	Comparative content	Traditional finance	Intelligent finance
1	Business philosophy	Product-centric	Customer-centric
2	Information processing	Difficult/high cost	Easy/low cost
3	Risk evaluation	Information asymmetry	Data is rich and complete, information symmetry
4	Capital supply and demand	Match through mediation, term and quantity	Solve by oneself completely
5	Payment	Payment via bank	The unity of super centralized payment system and individual mobile payment
6	The Parties of supply and demand	Indirect transactions	Direct transactions
7	Cost	Very high transaction cost	Lower transaction cost

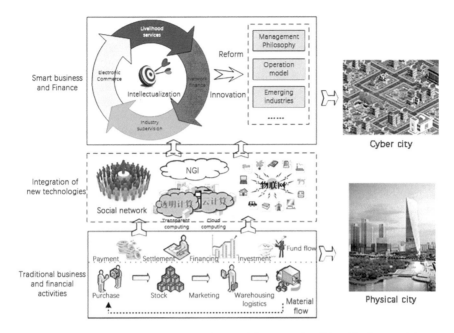

Fig. A.8 Difference between intelligent business and finance and traditional business and finance

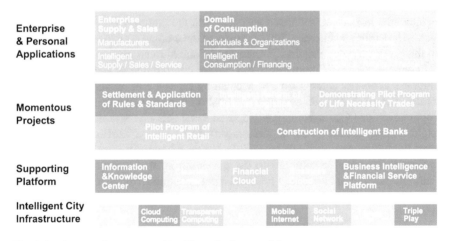

Fig. A.9 Construction content of intelligent business and finance

used mainly include statistics, forecasting and other operations research methods; customer management, supply chain management, enterprise resource planning and other management theory and methods; and enterprise modeling methods, etc. See Fig. A.9 for details.

The Developmental Status of Intelligent Business and Finance and Vision

1. The application developmental status of foreign intelligent business and finance

With the development of business and finance, its intelligence is one of the future trends. At present, there is about 1000 projects of intelligent city have started or under construction around the world, in which the intelligent construction of business and finance is the important components.

For example, the intelligent city in Europe pays more concerns on the effect of information and communication technologies on city's ecological environment, transportation, medical, intelligent buildings and other fields of livelihood, expecting through the knowledge sharing and low carbon strategy to achieve the emission reduction targets, to promote the city's low carbon, greening, sustainable development, and the construction of green intelligent city.

The London municipal government has successively proposed the e-London and London Connects strategies, both of which are based on how the government can provide better public services, illustrating the important role of intelligent business in delivering better services and reducing cost.

In April 1997, the European Union introduced the "European e-commerce initiative", and in February 1999 proposed the international charter which aims at coordinating the global communication, especially e-commerce. Since then, the e-commerce in Europe developed very fast. But the domestic e-commerce level of each EU member is different, 70% turnover of the e-commerce market concentrates on three key markets: Britain, Germany and France. According to the statistics by EU Communications Commission, the current contribution of e-commerce to the European economy is less than 3%, accounting for 3–4% of retail sales. In order to achieve the EU's goal of doubling the total amount of e-commerce in 2015, the Communications Commission has proposed some action programs such as rapidly implementing the European intellectual property strategy, further developing the bank card payment, network or mobile phone payment market, consulting the public for the possible solutions to the issues during the package delivery, especially the international delivery business, strengthening training for online traders, to let them understand their obligations and opportunities in the digital single market.

E-commerce originated in United States, and the highly developed market economy system provided the good economic, technical and social conditions. Therefore, the e-commerce development in United States always maintains a global leading position. The development of e-commerce in United States mainly concentrates on two fields: B2C e-commerce and B2B e-commerce. The facts show that the application of networks and e-commerce has produced the immeasurable economic and social benefits.

The rapid development of e-commerce in United States, to a large extent thanks to the government's strong support. Once e-commerce appeared, the US federal and state governments had great enthusiasm for its development, and developed lots of relevant laws and regulations and policies to promote the e-commerce development. In addition, the free super WIFI network across the country created by US government provided a good information network infrastructure for the e-commerce development.

In the financial field, Citibank and the United States M1 company in January 1999 jointly launched a mobile banking; customers can use GSM mobile banking to know the account balance and payment information, and send text messages to the bank to perform transaction. Additionally, customers can also download personalized menu from Citibank, and read the notice from the bank and check the financial information. US telecom giant AT and T launched a mobile banking business through the Cingular Wireless business model, jointed with four banks, making the mobile phone into a credit card. It is expected that by the end of 2015, the scale of US mobile banking users will reach 60 million people.

In 2004, Japan proposed the "U-Japan" program, aiming at building Japan into an environment where "any time, any place, any thing" can access to the internet. In 2009, the Japanese government also proposed the "i-Japan (Intelligent Japan) 2015 strategy". This strategy tried to develop three core fields including the intelligent transportation system and high degree logistics system, the e-government development, medical and education, which aimed at by 2015 achieving the people-oriented, "peace of mind and full of vitality, digital society", and thus promoting the whole economic society, giving birth to new vitality, and achieving the active and independent innovation. The Japanese government emphasized that the intelligent city construction focused on the people's livelihood, so that the citizen could see the real interests, and make full sense of the construction. The enthusiasm of local government to participate in the intelligent city construction was particularly high.

From the perspective of improving efficiency and reducing costs, many Japanese companies have improved the production efficiency, business management, inventory management, customer management and online purchasing, etc through e-commerce. The effect was very significant, generally, the cost reduced by 10–30%. Japan's e-commerce is characterized by: (1) Mobile Internet becomes a feature; (2) convenience store (Combines) is in the ascendant; (3) a variety of payment methods coexist.

In 1992, Singapore proposed the IT2000—intelligent island program, planning to build the high-speed broadband multimedia network across the country within 10 years. In June 2006 it also announced the "Intelligent Country 2015 (IN2015)" program.

To ensure the smooth implementation of the program's objectives, the Singapore government has specifically determined four key strategies: building a new generation of information and communication infrastructure, developing the information and communication industry with global competitiveness, developing the information and communication human resources which is proficient in information and communication and with global competitiveness, achieving the transformation of key economic fields, government and society.

2. The application developmental status of domestic intelligent business and finance

The intelligent construction and application in business and finance field have achieved rapid development. The effects in Beijing, Wuhan, Ningbo, Xiamen, Guangzhou, etc. are more significant. For example, Beijing issued the "Intelligent Beijing Platform for Action", putting forward the Electronic Toll Collection (ETC), "electronic green standard" and other intelligent applications; promoting the "citizen card" (including the social security card and real name traffic card, etc.), so that the citizen could enjoy the medical, employment, pension, consumption payments and other social services with the card; promoting e-commerce applications, etc. Wuhan proposed the 15 key fields with the most urgent needs, which included: the intelligent tourism, intelligent education, intelligent water, intelligent food and drug safety, intelligent community, intelligent logistics, and intelligent space, etc. Ningbo, the city that first systematically deployed the construction of intelligent city in China, clearly put forward the "Six Speeding Up" major strategy, the construction of intellectual logistics, intelligent health security, intelligent credit management and other projects. The intelligent financial parts of the wireless urban agglomeration construction in Fujian mainly include wireless POS application system, mobile bankcard consumption, mobile payment, account management, financial escort personnel identity identification transfer system, wireless sales management terminal, which provides a better platform for the convenient consumption and expanding consumption ways. The online payment, government informatization, intelligent logistics, enterprise informatization and life informatization and other aspects of intelligent applications are in-depth in Xiamen City.

11 fields such as intelligent logistics and commerce will be on the top of the agenda of Guangdong. For i-business, Guangdong proposed to establish an intelligent business support system to support the establishment of Guangdong international e-commerce credit platform. It promotedinter-industrial cooperation between banks and enterprises, and built a safe, fast and convenient online payment management platform, and developed online payment value-added services.

New technology is widely utilized in trade and finance sector. Intelligent commerce and finance shall be promoted with rapid advance of information technology. We will make the most of internet and computer technology to drive forward new industries. Innovation and progress in this form are made in internet, cloud computing and online payment.

Emerging business modes keep cropping up. For example, e-commerce has experienced explosive growth. In China, e-commerce revenue quadrupled in about 5 years. Online retailing had an average growth rate of 80% for the past 5 years. Since 2003, annual compound growth rate has been 120%. In 2012, China recorded its revenue of 8000 billion yuan in e-commerce, only next to the U.S. In 2013, China recorded its revenue of over 10,000 billion yuan in e-commerce, ranking first in the world (Ruan 2014).

Internet finance has become an important force to promote China's financial reform. Internet finance is an emerging sector that integrates traditional finance and internet. It refers to a rising finance of integrated fund, payment and information agent with the help of internet and mobile communication. It is different from indirect financing among commercial banks, and from direct financing in capital markets. It consists of 3 basic forms of enterprise organization: onlinemicro-credit company, Third Party Payment Company and financial intermediary company.

3. **The development and vision of intelligent business and finance**

(1) Extensive interoperability, comprehensive perception. Based on the internet, internet of things, telecommunications networks, radio and television networks, wireless broadband networks and other network combinations, using a new generation of internet of things, cloud computing, decision analysis and optimization technology, etc, through the instrumented, interconnected, intelligent ways, to connect the physical infrastructure, information infrastructure, social infrastructure, urban information resources and business infrastructure together from city-to-countryside, thus to achieve the more intelligent business and financial activities through perception and interconnection.

(2) Business integration, city and countryside unbounded. The development of intelligent business and finance will gradually fade the business activities of manufacturing enterprise, the traditional procurement, sales, logistics will be uniformly undertook by the professional company, to achieve a more comprehensive and professional division of labor; financial accommodation and payment will be gradually be undertook by other financial institutions or financial patterns, other than the traditional banks. The intelligent business and financial activities can make the boundary of city and countryside more and more blurred, narrow the differences between city and countryside, boost the urbanization and urban and rural integration, so that people's life and work in urban, suburbs, and even rural areas will gradually converge with the same convenience, easiness and high efficiency.

(3) Intelligent decision-making, full service. Including the intelligent decision-making control, whole course of e-commerce, one-stop financial services. All professional services of financial nodes rely on the open service platform, interconnecting, interchanging information are for the formation of close division of labor and cooperative relationship, and for the formation of a complete one-stop service package presented to the users.

(4) Transparent transaction, "visual" process. (1) Customers complete the transaction in a more transparent environment, and obtain a wealth of information from an unprecedented extensive source. (2) Full visibility into the supply, inventory and sales channels. (3) Internet of things and other new technologies achieve the intelligent logistics operations, networking and automation, so that the entire logistics supply chain becomes more transparent and efficient.

(5) Easy life, anytime and anywhere. (1) Customer-centric. Through the initiative, interaction, user care and other multi-angles to communicate with the users, ensure timely and appropriately to provide the products or

services with reasonable prices and in line with customer demand, as well as more convenient payment settlement, more timely financial services, and more secure wealth management. (2) Transaction at anytime, and delivery at anywhere. The purchase place is no longer limited to shopping malls; the purchase time is no longer limited to the business period; and the formation of business decision-making is no longer limited to meeting discussion.

Overall Strategy on Construction and Promotion for Intelligent Business and Finance of Our Country

1. Guiding Ideology

Adopting the Scientific Outlook on Development as guide and adapting to the requirements of changing mode of economic growth and building well-off society in an all-around way. Taking enterprises as the mainstay, continuously innovating and deepening the application modes while facilitating the role of government to strengthen the sharing of resources. The informatization and intellectualization degree in business and financial field should be further upgraded with the help of major public infrastructures, service platform supporting and application project pilot demonstrations to enhance the developmental level of "Four Modernizations Synchronization" national strategy.

2. Basic Principles

(1) Demand as traction and enterprises as mainstay. Making full use of the dominating function of relevant enterprises in intelligent business and finance development. Sticking to the principle of market-orientation and making full use of the decisive role for market mechanism in resource allocation. Exploring for low-cost and effective business development modes.

(2) Resource sharing and government promotion. For public information resources, on the basis of security, the phenomenon of information barriers and information isolated island should be broken mainly by using the power of government administration and law; For industry information and enterprise information, the government should promote the legal system construction in information security, information standards and other fields to eliminate obstacles for full interconnection and information sharing.

(3) Innovating the development and deepening the application. Sticking to pragmatic innovation, selecting the correct entry point, and focusing on the applicability and effectiveness. We should be people-oriented and benefit all the people. We should concern about the people's livelihood to create available, affordable and user-friendly application forms for the broad masses. We should strive to achieve positive interactions between applications and network, technology, as well as the industry.

(4) Planning as a whole and proceeding in order. The framework of intelligent business and financial cannot be divorced from the reality of the city. It should be proceed in a phased and orderly manner according to urban development strategies, technology evolution trends and social livelihood needs for different periods.

3. Strategic Positioning

(1) Intelligent modern service industry becomes a booster for economic and social development

Business and financial services will be further expanded, the service mode will be more diverse, and the quality of service will be more high-quality, convenient, safe and efficient. The potential for economic transition, value creation and social change will be further apparent.

(2) Important strategic emerging industries become a new economic growth point

Intelligent business and financial industry will have flexible and diverse forms, fast-growing industrial scale, and will become important strategic emerging industries of our country. It will further play its leading role in the tertiary industry, boosting the rapid developments for information technology, logistics, creative industries and other related industries, and also the rapid expansion of domestic market demand. It will become a new economic growth point.

(3) The main production-element allocating market becomes the economic base of intelligent city development

It has achieved the interaction for material flow, capital flow, information flow and optimal allocation in city and promoted the rapid development of urban economy; At the same time, as the leading industry in the service industry, intelligent business and finance will promote the development of integrated urban and rural convenience service system, and they will also lead to the changes in living and working modes and boost the development for related services. All these are conducive to change the space form for modern city, therefore the city development will skip the "urbanization trap" and achieve a more sustainable development goal.

4. Strategic Goals

The overall goal: the intelligent level in business and financial field become further enhanced, the infrastructure and service support platform become gradually improved, and realizing a full sharing and comprehensive deep-utilization for relevant information resources; The business and financial applications should be further innovated and deepened, and its contribution to national economy and social development should be significantly improved. The proportion of intelligent business and financial activities in the modern service industry should increase significantly. The security system for intelligent business and finance should be sound, stable and reliable.

Specific goals for the 12th five-year plan period include:

(1) Building up a number of major infrastructure platforms and integrated service platforms. Building up a number of intelligent business and financial infrastructure platforms with sufficient capacity and stable operation to achieve a comprehensive perception and interconnection; Achieving a full sharing and knowledge discovery for business infrastructure, information infrastructure, social infrastructure and urban information resources.

(2) Building up a number of demonstration applications for key industries. Constantly expanding the application of intelligent business and finance in various fields such as industry, agriculture, commerce and trade, transportation, finance, tourism and urban and rural consumption. Promoting the pilot application of intelligent mobile business in various fields such as agricultural production and circulation, business management, safety in production, environmental monitoring, logistics and tourism services. Gradually carrying out the construction of intelligent banking, intelligent securities and intelligent insurance to achieve the network and mobile operation for banking, securities and insurance business; Exploring new financial services such as mobile payments, mobile banking, customer centers, online banking, online securities and online insurance to create a new interactive and collaborative mode for financial clients, employees, managers and regulators.

(3) Cultivating a number of operational service enterprises and financial institutions with innovative business modes. On one hand, promoting the enterprises' e-commerce to develop in the direction of business integration and collaboration such as R&D design, manufacturing, operation and management. On the other hand, speeding up the improvement of intelligent level in new service industry such as network value-added services, e-finance, modern logistics, service outsourcing, chain operations, professional information services and consulting intermediaries. At the same time, cultivating a number of research and development enterprises with independent core technology in intelligent business and financial field and forming the industrial scale.

(4) Develop a great number of talents that adapt to intelligent business and finance building; work to develop an efficient and applicable personnel training system; provide adequate knowledge to high-end talents with professional skills in urgent need in intelligent business and finance circles; encourage industrial organizations, professional training institutes and enterprises to provide adequate knowledge to intelligent business and finance talents so that they better perform their jobs.

Top of the Agenda

1. Encourage intelligent business and financial supporting

We need to focus on supporting strong enterprises to establish 5–8 intelligent e-marts and financial platforms that are of international influence, and to develop intelligent business centers.

(1) We need to guide to establish a unified data platform, so that data can be shared among regional application systems of multi-level business and financial information systems.
(2) Support the establishment of a big data processing center to contribute to development of intelligent business and finance.
(3) Support the establishment of cloud business and cloud finance; encourage enterprises to establish private clouds in business and finance sectors, so that resources can be used on demand and charged according to use.
(4) Promote intelligent business centers. Development of intelligent business center is a unified system that involves information infrastructure, access network, internet of things, e-commerce, shopping guide, information collection and distribution, warehousing logistics and security monitoring that provides services for and underpins customers, businesses and managers.

2. Promote intelligence of key industries

Promoting intelligence of key industries such as modern logistics, e-commerce, retail, banking, securities and insurance is on the top of the agenda of intelligent business and financial development strategy.

3. Boosting the Intelligent Process of Large-scale Enterprises

Encourage the qualified large-scale enterprises and financial institutions to transform the e-commerce platform into industry intelligent business or intelligent financial platform. Making full use of the demonstration and leading role of large-scale enterprises with an early start point in informatization construction and an abundant accumulation of informatization such as telecommunications, banking, insurance, medical, tobacco, manufacturing and retail. Effectively promoting other enterprises in the industry to accept and promote intelligent business and financial applications. Deepening intelligent business or intelligent financial applications of the large-scale enterprises, improving the level of business or financial services, improving the operational efficiency and expanding the distribution channels and market space. Encourage the qualified large-scale enterprises and financial institutions to transform the e-commerce platform into industry intelligent business or intelligent financial platform.

4. Improving the Efficiency of Service for Intelligent Business and Intelligent Financial Institutions

The improvement for efficiency of service for intelligent business and intelligent financial institutions mainly include the following three aspects:

(1) More efficient. It mainly affects the efficiency of business and financial institutions through three aspects—more thorough sense of measurement, more comprehensive interconnection and deeper intelligent insight, so that customers can access their desired services and a variety of personalized services anywhere, moreover, it will also greatly reduce the agency's labor costs and improve work and service efficiency.

(2) More comprehensive. First of all, by conducting data collection of enterprise's industrial chain and using technologies such as cloud computing to manage and control the fund demand for each link of an industry chain and the personalized services required to be provided, helping relevant staff to access required information and applications from different service channels in a convenient manner, and discovering potential businesses and cross-selling organizations to serve customers in a comprehensive manner. Secondly, intelligent business or intelligent finance can also help customers to switch between a number of service channels seamlessly in transaction process, reducing duplicate work and providing customers with a more perfect service experience. Thirdly, business or financial institutions can also provide a number of risk control tools. According to their own operating conditions and environmental changes, they can provide the risk assessment results at any time, and issue warnings at a critical time to remind enterprises for counter-measures, thus helping enterprises to control and resolve risks.

(3) More convenient. By using various types of intelligent information platforms, business or financial institutions can also provide customers with more convenient services, so that customers are able to enjoy the services meeting their own requirements anywhere. Efficient network technology allows providing customers with tailor-made services in the shortest possible time.

5. Strengthening Business and Financial Big Data Mining and Comprehensive Development and Utilization

Achieving the deep development of business and financial information resources, as well as the integration and utilization of information resources in fields such as manufacturing, medical, education and others. Basically meeting the information needs of various fields for intelligent cities, and promoting the transformation of economic growth mode and the construction of resource-saving society. Speeding up the construction of e-commerce cloud and financial cloud, managing and digging the huge value of unprocessed loose data in social, mobile networks, e-commerce and financial institutions large IT facilities. Encouraging the enterprises to innovate technologies and participate in processing and analyzing big data.

Expanding the relevant application services to meet various requirements in national and local economic construction. Guiding and standardizing the social value-added development and utilization for business and financial information resources which serving the intelligent city.

Security System for Strategy Implementation, Measures and Recommendations

1. Basic Framework of Security System for Strategy Implementation

First of all, the process of industrial development of intelligent business and finance shows obvious features of industrial development for continuous technological innovation. Scientific and technological innovation resources with the core of meeting high-level professional and technical personnel's, and the multi-level capital supply market adapting to the fund demand for high-tech industry are the most important factor endowment guarantees rather than natural resources, land and general labor force.

Secondly, the development of financial ecological environment and market cultivation mechanism required by the future-oriented business and financial development is seriously lagging behind. Building up a financial ecological environment with the healthy operation for financial system as core and an industrial development policy with the perfect market cultivation mechanism as core thus become the most critical "software infrastructure" guarantee contents for achieving the intelligent business and financial development strategic goals.

Thirdly, the intelligent business and financial development strategy has significant features of synergistically developed economic relation. The most critical point is to strengthen the organization and coordination role of the government in the process of industrial development. This is the most important organizational guarantee in the implementation process of the intelligent business and financial development strategy.

Fourthly, a perfect and effective intelligent business and financial development evaluation system is the decision-making guarantee for dynamic adjustment during the implementation process of the strategy.

Therefore, the basic structure for promoting and implementing security system for intelligent business and financial development strategy should be built from six most important functional dimensions (as shown in Fig. A.10).

2. Measures and Recommendations on Security for Strategy Implementation

(1) Strengthening the organizing and cooperation of government, and comprehensively enhance the function of coordination service. Strengthening the longitudinal cooperative relationship between leading organization departments at all levels for informatization work; Strengthening the lateral collaborative

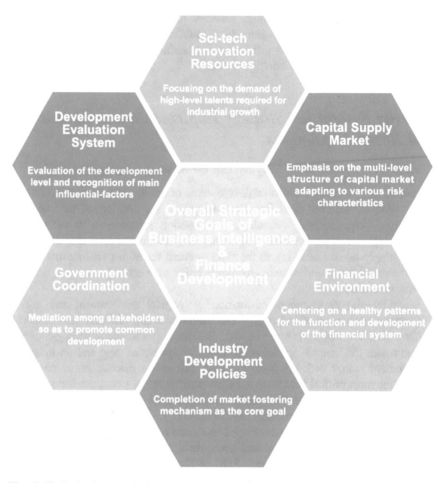

Fig. A.10 Basic framework for implementing security system for intelligent business and financial development strategy

relationship between leading organization departments and competent departments for industrial development.

(2) Improving the industry development policy, and promoting the optimal
allocation of social resources. Building up a comprehensive policy supporting
system for intelligent business and financial industry development;
Strengthening the compatibility of various types of industrial development
policies; Focusing on the improvement of actual effectiveness for each policy
during the implementation process.

(3) Optimizing the financial ecological environment to ensure the healthy operation
of the financial system. Improving the external regulatory system to maintain
the safe and stable operation of the financial system; Deepening the reform of
financial system and eliminating the institutional defects of financial property

rights; Strengthening the construction of credit system and credit institution in the whole society.

(4) Accelerating the construction of infrastructure and establishing an unified information management platform. Building up a multi-level information network platform with an integration of information integrity and security; Strengthening the construction of basic database; Actively exploring the construction of products cloud, logistics cloud and financial cloud to promote the extensive use for new generation of information technology in intelligent business and financial field.

(5) Improving the value creation mode and making use of the function for variety of business entities. Making full use of the leading role of operators in industrial chain and the supporting role of solution providers.

Abstract of 'Infrastructure of Temporal-Spatial Information of Intelligent City
A Development Strategic Research in China'

Geospatial information infrastructure and network infrastructure, perceptual infrastructure, cloud computing center infrastructure together constitute the infrastructure for intelligent city.

Geographic Information Infrastructure for Intelligent City

Geospatial information infrastructure of the intelligent city mainly includes space-time datum, space-time data resources, unified information resource management and service platform, support system and so on.

Space-time datum provides target positioning, data fusion and multi-sensor integrated datum framework; Unified information resource management and service platform provides integration, management and networking services for mass, multi-source and heterogeneous city data; Space-time data resources mainly include basic geospatial data and dynamic sensing space-time data, the latter one includes dynamic incremental data such as moving target data, monitoring data, urban residents' POI data and emergency management data; Support system includes technical system, policy and standard system.

Construction Goals and Ideas

1. Construction Goals

On the basis of the digital city geospatial framework, we should take the advanced technology (with the new generation of information technology as core) as driving force and take the comprehensive perception, comprehensive interconnection, deep integration, collaborative operation and intelligent service for the city system in the highly-informationized era as goals. We should build up an ubiquitous real-time service system of dynamic acquisition, integration, processing, decision-making and service for urban real-time spatial data flow and achieve an ubiquitous, real-time and intelligent information services on the basis of digital city to promote economic, living, social and city management into the intelligent development stage.

2. Construction Ideas

According to the "basic, public and industrial" characteristics of the geographic information infrastructure for intelligent city, establishing that geographic information infrastructure is an unified and authoritative public information platform for loading and distribution, management release, statistical analysis and decision-making services of all kinds of information from the national level. Establishing a three-in-one combination development mode with top-level design controlled by the nation, basic investment taken by the government and service requirements guided by industry and society to change the construction and development mode which relies solely on the national investment. Promoting the formation of independent development ability for construction of the geographic information infrastructure with the geographic Information Industry.

With the space-time information and dynamic update as core, enriching, expanding and upgrading the digital city geospatial framework. Building up a multi-source, multi-scale and unified dynamic geospatial framework with the help of the network infrastructure, perceptual infrastructure, cloud computing center infrastructure of the intelligent city.

On the basis of the digital city geospatial framework, promoting the "space datum" into "space-time datum", the "two-dimensional geographic information + three-dimensional visual expression" into "four-dimensional geographic information for unified space-time datum", and the "post-analysis + auxiliary decision" into "real-time analysis + real-time decision-making"; Strengthening the construction of the geographic information infrastructure including space-time datum system, geospatial information integration management platform, intelligent perception, access and fusion platform, intelligent analysis, decision-making and service platform; Organizing the power to carry out scientific research on core technologies such as the generic specification coding framework that describes various types of sensors and their interrelations, real-time access and correlation of multi-source sensor information under the unified time and space, self-loading and content fusion of multi-source and heterogeneous information, and variation-oriented efficient information updates. Establishing the supporting laws and regulations, specification standard system, and promoting the standardized development of geographic information infrastructure construction.

Construction Content

Integrating the geographic information resources required by government departments, enterprise and public institutions of the public within the city in accordance with unified standards. Building up a distributed database with the basic geographic information database as framework to achieve the interconnection and information resources sharing for multi-level integrated of geospatial information resource service platform on national, provincial, municipal and district (county) level. Developing network system and integrated service platform for data releasing, sharing, exchanging and service. Providing the government departments, enterprises and the public with authoritative, accurate and strong-currency geospatial information services and functional services through online service. Building up a cross-sectorial, cross-industry, cross-network, cross-platform, high service aggregation, high reusability, high availability and low threshold management system and operating mechanism for sharing, exchanging and updating, as well as the related standard specifications and security supporting system.

1. **Building up the Geospatial Information Integration Management Platform and Forming the Intelligent City Data Center**

On the basic of basic geographic information database for digital city, enriching the multi-temporal basic geographic information data and new product data such as panorama image and point cloud. For the entity object data, adding time attributes and forming the space-time information data set. Obtaining and standardizing the name and location of the nodes for Web of Things, and unifying the classification and coding to form static and dynamic Web of Things node address data set such as IP address and QR code. Building up the space-time information database; Achieving various types of data storage including two-dimensional, three-dimensional and metadata information with the improvements on functions and integrated management for the mass heterogeneous multi-source spatial data set of the data center to provide an intuitive display platform for intelligent city and provide the basis and support for instrumentation, interconnection and intelligentization.

Building up an unified data resource sharing and exchanging platform and information resources exchanging system. Achieving the information sharing and exchange including various types of data resources, media resources and business resources of the city. Building up an unified service registry, global service directory, operation and maintenance control for resource status and authorized certification management to fully realize an unified exchange management, unified authorization management and unified operation and maintenance control; Building up a mass intelligent city space-time data warehouse and laying the foundation for online analysis of city information, including: city multi-source heterogeneous space-time data representation model, city space-time datum and address coding method, rapid integration method for multi-source heterogeneous information, management method for city infrastructure and component monitoring information, method for detecting the rapid changes in space-time data real-time updating, space-time data compression and multi-dimensional visualization method.

2. Dynamic Information Intelligent Perception, Access and Fusion Platform

Building up a multi-source data integration framework system of "geospatial information + sensor network + Web of Things" to form a city information perception and access path with ground, surface and underground, static and dynamic, indoor and outdoor in combination, comprehensive coverage and complement each other. Building up a new generation of positioning service system adapting to variety of satellites, multi-frequency and multi-mode signal to reflect the state of the city from multi-angle. For the requirements of achieving interconnection, intelligibility and interoperability for multi-source heterogeneous sensors of the city: Building up the generic specification coding framework that describes various types of sensors and their interrelations. Achieving a real-time access and correlation of multi-source sensor information under the unified time and space, self-loading and content fusion of multi-source and heterogeneous information, and the variation-oriented efficient information updates. Building up a distributed, multi-source and heterogeneous information management platform for real-time accessing multi-source information, self-loading heterogeneous information and orienting the intelligent city. Conducting integrated organization and processing on space-time information to form a spatial information infrastructure with on-demand service capability, powerful spatial data management and information processing capabilities.

3. Basic Geographic Information Cloud Service Platform

Digital city geospatial framework supports the environment to expand into "cloud mode". It should be transformed into "cloud" mode in aspects such as hardware and software infrastructure, public service functions and system data. Reducing capital investment, shortening construction cycle and providing better service supports through this mode. Supporting the rapid construction of intelligent city business application system. On the basis of the building configuration development technology provided by various information service platforms in the information service network, system developers can call, mix, assemble, and process various information services deployed on the network and build application system that meets business requirements in a rapid way. The key items to be constructed under the cloud architecture are:

(1) Intelligent analysis platform. Carrying out data mining and knowledge discovery based on multi-source, multi-temporal (aboveground and underground, indoor and outdoor) geospatial information and various types of sensor data

accessed in real-time. Building up the expert knowledge base for different topics and fields through research works such as real-time process simulation method, geospatial statistical method, clustering method, correlation analysis, classification and forecast analysis. Exploring the potential and valuable information, rules and knowledge of the space-time system from geospatial database. Reflecting the interrelation and the dynamic changes of spatial elements in real-time and monitoring urban anomalies in real-time by using geographic computing models and methods and real-time analysis technology for dynamic information. Improving the intelligent level for geographic information applications by updating geospatial data using video information to provide powerful tools for researching and solving complex problems.

(2) Intelligent decision-making platform. Integrating the intelligent analysis and modeling method of heterogeneous data, building up intelligent services, decision support and collaboration platform with the help of information resources integrated management platform and general information integration platform to provide the decision support capability for multi-level and multi-user and to lay the foundation for city operation and management, construction planning, emergency command and decision support. The main works include: intelligent mining analysis for multi-source data, intelligent spatial analysis and statistics, intelligent video retrieval analysis, analysis and decision model simulation, domain knowledge base construction and auxiliary decision-making and decision-making model service chain intelligent combination.

(3) Intelligent service platform. Following the ideas of "open standards—service encapsulation—service registration—service portfolio" and extracting business processes and functional requirements for industry applications. Splitting off some of the core functions from the processes that are closely related to the industry and packaging them into a general service for various industries. Providing external service interface and the applications for demonstration industries.

4. Security System

Building up a cross-sectorial, cross-industry, cross-network, cross-platform, high service aggregation, high reusability, high availability and low application development technology threshold (four-across, three-high and one-low) management system and operating mechanism for sharing, exchanging and updating, as well as the related standard specifications and security supporting system.

Promotion Strategy

1. **Promotion Route**

Adopting a three-in-one combination development mode with "top-level design controlled by the nation, basic investment taken by the government and service requirements guided by industry and society". Promoting the strategy and implementing the geospatial infrastructure construction in accordance with the "overall planning, phased implementation, information sharing, comprehensive promotion" principle.

(1) Strengthening the integration for existing geospatial infrastructure (datum, framework data, databases, geographic information systems), building up a common basic database and integrating the information infrastructure of the whole city. Developing standard specifications, evaluation indicators and checkup system that supporting the intelligent city geospatial infrastructure.

(2) Conducting the upgrade focusing on key points such as geospatial data resources, platforms and systems and updating the construction. The focus is to eliminate the information isolated island, to establish the spatial information sharing and exchange platform, to achieve information sharing and the interconnection, intelligibility, interoperability between systems.

(3) Fully implementing the content for intelligent city geospatial infrastructure construction. Building up the intelligent city geospatial infrastructure that can fully support the intelligent city with integrated and intelligent geospatial information collection, processing, storage, management and service capabilities.

2. **Breaking Through Key Technologies**

For the problems in interconnection, intelligibility and interoperability for multi-source heterogeneous sensor information of the intelligent city: Breaking through the core technologies such as generic specification coding framework that describes various types of sensors and their interrelations, real-time access and correlation of multi-source sensor information under the unified time and space, self-loading and content fusion of multi-source and heterogeneous information, and the variation-oriented efficient information updates. Building up a distributed, multi-source and heterogeneous information management platform for real-time accessing multi-source information, self-loading heterogeneous information and orienting the intelligent city. Conducting integrated organization and processing on space-time information to form a spatial information infrastructure with on-demand service capability, powerful spatial data management and information processing capabilities.

(1) Organization, management and analytical application for space-time data, including distributed integration and sharing, dynamic integration and

space-time data assimilation; The support software's include (grid) geographic information system, remote sensing information system, three-dimensional visualization software, distributed database system and so on.

(2) The existing GIS are difficult to meet the real-time access, fusion, processing and efficient dynamic update management for multi-source heterogeneous sensor information. It is urgent to make key technological breakthrough on GIS for multi-source sensor information.

(3) The existing geographic information sharing and interoperability technology has not systematically considered about the real-time observation information of multi-source heterogeneous sensors. It is necessary to break through the real-time access and correlation technologies for sensor information under the unified space-time system. Unified access and loading standards are required to build multi-source heterogeneous sensor information.

(4) In order to form the subject-oriented comprehensive information, the ability to load and fuse multi-source heterogeneous information is required.

(5) For the mass sensor information, it is required to provide the data update mechanism basing on abnormal changes discovery.

(6) Achieving the spatial information intelligent processing for city operation. It is necessary to establish the real-time access, dynamic loading and comprehensive integration platform for space-time information that supports city comprehensive management.

3. Policy and Legal Protection

The construction content for policy and legal protection for intelligent city surveying and mapping geographic information infrastructure mainly includes:

(1) Policy level. Carrying out a comprehensive review of the existing relevant policies for intelligent city surveying and mapping geographic information infrastructure and carrying out the research on policy mechanism. Submitting the relevant reports and suggestions to the State Council by defining the development orientation, key areas and security mechanism for surveying and mapping geographic information infrastructure. Draw a comprehensive deployment of the construction for intelligent city surveying and mapping geographic information infrastructure and introducing relevant policies as the basis for work development and promotion.

(2) Legal level. Clarifying the development conditions and boundaries for the existing relevant laws and regulations with reference to the legal system of the digital city geospatial framework and the construction demand for mapping geographic information infrastructure required by intelligent city. Carrying out research on the currently imperative issues that have no foundation and establishing relevant committees. Putting forward corresponding proposals on laws and regulations in the National People's Congress, and improving the relevant legislation for intelligent city surveying and mapping geographic information infrastructure.

(3) Planning level. Incorporating the "intelligent city" surveying and mapping geographic information infrastructure construction into planning at all levels of government. For most of the cities in China, the conditions for the transition from the information city, digital city to the intelligent city are not yet fully ripe. However, cities with good conditions mainly focus on the aspects such as communications infrastructure and computing infrastructure, they have no idea on the construction and application of surveying and mapping geographic information infrastructure. Therefore, with respect to the planning, we should try our best for the relevant policy support and protection, regulate the construction work relying on laws and regulations, and incorporate the intelligent city geographic information infrastructure construction into the 13th five-year plan at national, provincial and municipal levels.

(4) Specification level. At present, due to the specification standards are not uniform, information and application systems produced by various cities cannot dock with each other. Intelligent city and the construction for intelligent city unexpectedly form many new information isolated islands, methods and bases for the construction and application of surveying and mapping geographic information infrastructure are also varied. Therefore, establishing the basic standard specifications at national level and promoting them to various provinces and cities. Subordinate administrative departments and related institutions will set local standards in line with the actual condition on the basis of the specifications.

(5) Management system. As the information industry, communications industry, surveying and mapping industry are all involved in the intelligent city and intelligent city construction at present, it is required to determine the management agency and collaboration departments in the management mechanism, and define various functions to ensure the smooth development of the construction work.

Measures and Recommendations

1. **Establishing the Position of National Surveying and Mapping Basic Geographic Information Infrastructure to Prevent Repeated Construction and Achieve a High Degree of Sharing for Information.**

Geospatial information is the carrier of all kinds of information in intelligent city and the carrier for integrating all kinds of information. Building up an unified, authoritative national surveying and mapping basic geographic information platform. Establishing various information integration mechanisms and systems on the basis of geographic information and expediting new growth points of the geographical information industry.

2. Developing Geospatial Data Sharing Development Policy

Geospatial data is the driving force of the intelligent city development and triggers a great change in the productivity force and productive relationship. The era of data as the core competitiveness has arrived. National basic data, industry data and government data will form big data resources. The monopoly benefits of these resources are becoming increasingly prominent. The current situation of "emphasis on confidentiality and contempt for sharing" will form the barriers for big data value development and the industry. Developing antitrust and shared development policies for intelligent city geospatial data to break the industry monopoly and expedite new growth points of the information service mode and information industry of intelligent city.

Abstract of 'Intelligent City Evaluation System'

"Intelligent City Evaluation System" is one of the projects under the major strategic project "Strategic Research on Construction and Promotion of China's Intelligent Cities" set up by the Chinese Academy of Engineering, with academician Pan Yun he, the executive vice president of Chinese Academy of Engineering and chief scientist of the collaborative innovation center as the general project leader, and professor Wu Zhiqiang, the director of the collaborative innovation center as the project leader. This research project puts forward the evaluation indicator system which embodies the connotation of intelligent city by reviewing the status quo of multiple intelligent city evaluation system at home and abroad in a full and detailed way and combining the practices for promoting the intelligent city of various local governments and enterprises. It is expected that similar evaluation and comparison can be conducted with the same logical ideas and methods on cities in China and other areas of the world and to promote intelligent cities around the world to advance towards a more sustainable direction in common and in communication.

The evaluation system of the intelligent city is to evaluate and guide the city's intelligent development level, and it is required to embody the connotation of intelligent city:

(1) Intelligent city cannot have any informatization for infrastructure, and can not only have the informatization for infrastructure;

(2) Intelligent city promotes city development and innovation through the information technology infrastructure, thus it becomes an intensive, intelligent, green and low carbon thruster for the new-type urbanization path to promote the sustainable development in three major areas (society, economic and environment) of the city. Therefore, with respect to the evaluation system of the intelligent city, it cannot be divorced from the main indicators for three major areas (society, economic and environment) of the sustainable development city, and cannot evaluate the development criteria of intelligent city by using these indicators only;

(3) In the field of intelligent city development and construction, achieving a four-zation fusion development with informatization optimizing urbanization, informatization promoting industrialization and informatization promoting the coordinated development of agricultural modernization;

(4) The city's intelligent development cannot be divorced from the public's educational quality and intelligent creativity. People are the service object of intelligent city, and also the fundamental driving force for the development of intelligent city. For the evaluation criterion of the intelligent city, the citizens' cultural quality and intelligent innovation ability cannot be ignored.

Summary of Issues for Evaluation Criteria of China-EU Intelligent City

This research selected a number of domestic and international intelligent city evaluation system for analysis, including: "Pilot Intelligent City (District/Town) Certificated by Ministry of Housing and Urban-Rural Development of China", "Smart City Evaluation Indicator System by Ministry of Industry and Information Technology System", "GMTECH Smart City Development Level Evaluation", "China Wisdom Engineering Association Indicator Intelligent City (Town) Development Index", "Pudong Intelligent City Indicator System", "Nanjing Smart Nanjing Evaluation System", "Ningbo Smart City Development Evaluation", "TU WIEN Indicator System", "Int'l Digital Corporation Indicator System", "Intelligent Community Forum Indicator System", "IBM Indicator System" and "Ericsson Indicator System". On the basis of the research on the tertiary indicators of the evaluation system mentioned above, there are following problems in the analysis of the existing indicator systems:

(1) Emphasizing the informatization basis while ignoring the core meaning and overall sustainability of the intelligent city

In these evaluation systems, most of the evaluations for intelligent city or intelligent city assessments are emphasizing the construction of information infrastructure and the applications of new technologies in various city systems. So that can be said to be more concerned about the input rather than output. In fact, the goal of the intelligent city is the overall harmony and sustainability of the city. Informatization and technologies as tools and approaches should aim to serve the goal. In the case of achieving the corresponding goals, simplistic low-tech solutions should be preferred to. Evaluation systems with one-sided emphasis on the development of technology may cause problems such as technology stacking.

(2) Emphasizing the current state while ignoring the development trend

The measurement for the development of the intelligent city should include two aspects in time series: the current state and the future state. Intelligent cities should have expectations on sustainable development in the future. Indicators of the existing evaluation systems are basically concentrated on static evaluation for the current state of the city, and on quantifying and comparing the indicators for various systems of the city. The judgment on future development trend and change rate of the city is lacking. Even if the city is highly estimated in the static evaluation currently, if the city is in recession, it should be evaluated as a non-intelligent development trend and state.

(3) Emphasizing the comprehensiveness while ignoring the characteristics

For the systematic perspective of the city covering areas, the evaluation of intelligent city should cover various city sub-systems. The current evaluation systems, although their focuses are varied, are also developing in this direction. For the requirements of city development, we should focus on the cultivation for characteristics of city development. That is, not to unify the standards, but to accommodate more characteristics and possibilities for city development under the evaluation system with a certain range, and to guide the unique brand building for each city to form of a diversified intelligent city ecology.

(4) Unified standards or specific city oriented, lack of flexibility

The existing evaluation systems can be roughly divided into two categories by their evaluation objects: the evaluations of ministries or enterprises are usually for general object cities, districts and communities, and those of the local cities are tailored to the requirements of specific cities. The indicator systems of the former are too rigid, and they impose uniformity in all cases with a lack of full consideration of differences in the developments for different cities. The latter has a strong pertinence, thus it often runs well in a certain type of cities with the consistency of time series and the comparative significance, however, it may have no reference significance for other cities in the evaluation transplantation. The evaluation of intelligent city has a certain range of flexibility. The types and scales for diverse intelligent cities should be evaluated and guided to accommodate a greater adaptability.

(5) The acquisition and operation for indicator data are difficult, and lack international versatility

Some indicators of the existing evaluation systems such as "development and implementation plan for intelligent city", "organizations" and "policy and regulations" are too macroscopically in concept and biased towards the ideal and theoretical level. Their data are difficult to be quantified and collected, and cannot be pointed to a specific application, operation or promotion. Also, they don't have international versatility and universal applicability. All these are adverse to the comparison and diagnosis with other countries and areas of the world.

Development of Intelligent City Evaluation Indicator System

1. Method of developing i-city evaluation systems

According to definition of i-city building and problems of i-city evaluation standards, i-city building requires a unified i-city evaluation standard that can be approved and applicable nationwide. I-city evaluation system needs to take sustainable development as the very core. The system works to improve technologies and quality of people. It emphasizes trends of urban development process, and is approved and well recognized by international enterprises and academic circles in cities.

In the framework of intelligence and environment, intelligent management and service, intelligent economy and industry, intelligent hardware building and resident intelligent quality, there are 20 factors to be evaluated according to types (see Figs. A.11 and A.12). If these factors are similar or targets of intelligence are same or similar, those which are easier to get continuous data and those which with international comparison will be in the system. When it comes to getting factors to be evaluated, we do not get data from annual statistical bulletins, but from the internet in a more intelligent manner, so that evaluated results will be updated dynamically, urban development and city building will be analyzed in a more fast and intelligent manner. As primary indexes are developed, factors to be evaluated need to be collected and sorted out. In virtue of expert counseling with Delphi method, 56 questionnaires were delivered to academicians and experts at research groups in Chinese Academy of Engineering. Indexes were revised. There was increase or decrease in indexes as evaluation systems do not match with each other.

2. I-city case selection

The study delves into cases of development of a large number of i-cities.

According to the Ministry of Housing and Urban-Rural Development, in its national i-city piloting meeting on December 29, 2012, the first 90 pilot i-cities including 33 municipal cities, 25 i-districts, 20 counties, 3 towns and 9 parks were publicized. The research group, at first, tried to assess 33 municipal cities, especially 31 cities with media buzz from the first pilot i-cities. Target groups are possible to grow as i-cities.

31 cities will be comprehensively assessed with the method aforesaid. Ningbo, Wuhan, Wenzhou, Zhuhai and Taizhou ranked in the top five target cities.

3. Analysis ontest assessment of potential i-cities

(1) According to the study, cities such as Ningbo, Wuhan, Wenzhou, Zhuhai and Taizhou (with above 40) scored nearly five times as much as Liaoyuan and Lhasa (with 10–15). For pilot cities by Ministry of Construction and Ministry of Industry and Information Technology of the People's Republic of China, development of potential i-cities varies.

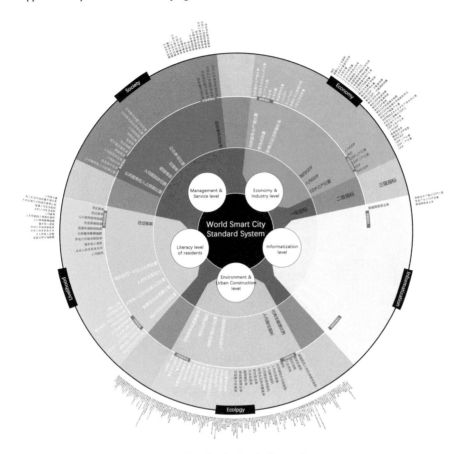

Fig. A.11 Framework of Intelligent City Evaluation Indicator System

(2) Top ranked cities scored 4 times as much as backward cities. It means 5 indexes in this study effectively assess i-city building, and in line with feelings of the masses on i-city building.

(3) Top ranked cities boast hardware infrastructure. In terms of hardware, Ningbo had 83 and ranked as the top i-city, nearly 80 times as much as poorly ranked cities that only had 1. It means, intelligent hardware building serves as leading reference in i-city ranking.

(4) For qualities of residents, top scorers such as Ya'an, Wenzhou, Wuhan and Zhuhai had over 50, nearly five times as much as backward players such as Liu panshui that scored 7.57. Polarization in scoring has been a priority for local authorities, i-quality of urban residents must be improved. Compared to intelligent hardware ranking, there is little difference in quality ranking.

(5) In terms of intelligent management and service as part of efforts of local authorities, top scorers such as Zhuhai and Dezhou had over 75, nearly six times as much as backward players such as Liaoyuan, Bengbu and Lhasa that

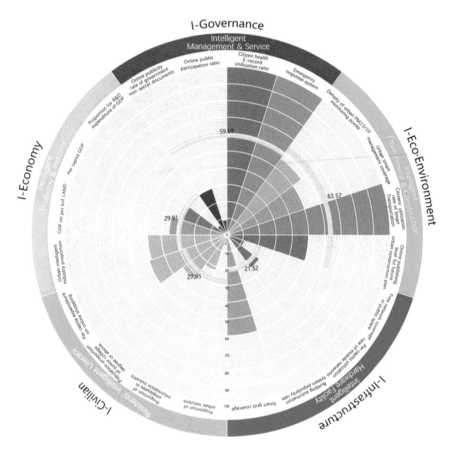

Fig. A.12 Intelligent City Evaluation Indicator System

scored less than 13. It suggests that a great number of cities still have a long way to go before they provide intelligent social services and intelligent government administration. I-city building quickens the pace to modernize social management.

(6) When it comes to intelligent economy and industrial building, besides Lhasa, top ranked cities scored nearly three times as much as backward rivals. It suggests that, thanks to reform and opening up since 1978, cities in developed and coastal regions or in inland China have been fully aware of intelligent technology's push for economic boom. Intelligent information technology that contributes to economic boom will be promoted fast in more cities.

(7) In terms of intelligent environment building, top scorers such as Jinhua, Taizhou and Ningbo had over 65, nearly five times as much as backward players such as Liupanshui and Lhasa that had less than 14. It suggests that, on the one hand, cities in developed regions invest money in environmental intelligence, on the other hand, as these cities are polluted, they fail to outdo

their rivals in intelligent environment building; on the contrary, some cities boasting environmental quality and coastal cities such as Jinhua and Taizhou with above-average scores are highly ranked.

(8) In general, the test evaluation helps to promote evaluation system. There is great headway compared to December 2013. It overcame socio-economic and environmental indexes without intelligent technology, and linked intelligence with socio-economic and environmental indexes that contribute to sustainable development for cities.

In addition, the study delves into leading cities in EU and the United States in i-city building. With these cities as evaluation references, we can fully know i-city building in China now and its development trends in a more objective manner.

I-city Evaluation Promotional Strategy

Strategy 1: Build integration with European countries and the outside world to develop a third-party authoritative evaluation system

In this study, we have collaborated with world-leading authorities such as Chinese Academy of Engineering, ACATECH and IVA, and promoted cooperation with i-city evaluation system. With joint urban development between China and EU, we learn from each other in terms of i-city evaluation. We will work to establish an authoritative third-party evaluation system that is well recognized in China and EU. Besides, as i-city evaluation is accessible to cities in China and EU, we work to promote the evaluation system.

Strategy 2: Work with Asian and American countries, build academic integration with universities and institutes

The study works to promote China's integration with Asian and American countries, especially have growing academic integration with developed countries such as Korea, Singapore and the United States. The study looks to build a long-term exchange and regular meeting system, and get its coverage of i-city development. The study strives to improve i-city evaluation system by building a growing talent pool. All that helps to promote i-city building.

Strategy 3: Publish i-city evaluation system for 2014 soon, establish an annual updating and reporting system to try and improve in practice

In this study, result of i-city evaluation system is tested. The system will be improved according to the tested cities. Follow-up "evaluation system", "evaluation updating system" and "evaluation promotion strategy" will be developed. An "annual report on i-city evaluation system" for 2014 will be published soon at the end of the year. Workable promotional suggestions on evaluated cities and alternatives will be developed.

Updated evaluation system depends on i-city declaration and evaluation. Indexes will be revised and updated according to development of target cities. We will remain committed to trying and improving the system.

Strategy 4: Accumulate information, dynamically evaluate, track and diagnose all i-cities

We will work to establish an i-city data pool nationwide that is accessible to all possible and potential i-cities. We must work to establish an early warning system for non-i-cities. All cities in data pool will be dynamically tracked and evaluated, i.e., independent and group evaluation. We must regularly publicize evaluation results, dynamic and real-time i-city ranking and diagnostic suggestion on i-city building. We will make evaluation system more instrumental and universal. We will accumulate information and practical experience for a better evaluation system.

Strategy 5: Strengthen close cooperation with local governments to promote i-city evaluation and its feedback

With evaluation system, talks with the local government must be promoted to put in place promotional partnership and co-build an exchange mechanism between local government and professional evaluation institutes. More frequent and stable evaluation system must be set up.

According to evaluations, we must identify and provide advice on weakness in i-city development, timely revise and adjust urban policies. At the same time, as the local government promotes the evaluation system, evaluation of i-city is data sourced, to have some feedback and improve the evaluation system.

Strategy 6: Build a public evaluation platform; get public response and supervision, in a move to get participation from below

I-city evaluation system is from the people and for the people. It emphasizes public feedback. A feedback system of evaluation and results is set up. Calculation and ranking of online publicity and evaluation system is set up. Professional analysis and public online information exchange system as well as online monitoring system are set up. We will work to depend on polling and establish a feedback system of i-city evaluation system from below.

At the same time, by promoting the evaluation system, we work to advocate i-city, its visibility and public awareness. Professional evaluators develop analysis reports on public participation and opinion feedback to deliver a result of annual i-city evaluation system.

References

European Commission. 2004. *MANUFUTURE—A Vision for 2020: Assuring the Future of Manufacturing in Europe*. Luxembourg: Office for Official Publications of the European Communities.

Fang, Z.X. 2011. Fuyang Plans to Build Smart City in Five Year. *Fuyang Daily*, July 22, 2011.

Gan, Q.C., J.C. Harreld, Y.W. Jiang, et al. 2009. *Smart Globe Wins in China*. IBM Institute for Business Value.

Luo, W. 2014. The enlightenment of Germany's industry 4.0 strategy to the industrial transformation and upgrading in China. *Industrial Economy Review* 4: 1–2.

Ruan, X.Q. 2014. E-business turnover exceeds 10 trillion yuan in 2013. *China Securities Journal*.

Shi, Y., and Y.S. Tan. 2012. Huzhou Carries out Pilot Site and Is to Create "Smart Industry" Framework in Three Years. April 04, 2012. http://zjnews.zjol.com.cn/05zjnews/system/2012/04/11/018404905.shtml.

Tong, M.R. 2010. "Smart City" construction: new opportunities for transformation and upgrading of manufacturing enterprises. *Ningbo Economy (Sanjiang Forum)* 11: 15–17, 21.

Xu, Y.M. 2012. Shunde Becomes the First Intelligent Manufacturing Pilot Site in China. *South News Daily*, Feb 20, 2012.

Printed in the United States
By Bookmasters